Electronics texts for engineers and scientists

Editors
H. Ahmed, *Reader in Microelectronics, Cavendish Laboratory, University of Cambridge*
P. J. Spreadbury, *Lecturer in Engineering, University of Cambridge*

Electronic circuit design: art and practice

Electronic Circuit Design:
Art and Practice

T.H.O'DELL

Reader Emeritus, University of London

CAMBRIDGE UNIVERSITY PRESS
Cambridge, New York, Melbourne, Madrid, Cape Town,
Singapore, São Paulo, Delhi, Tokyo, Mexico City

Cambridge University Press
The Edinburgh Building, Cambridge CB2 8RU, UK

Published in the United States of America by
Cambridge University Press, New York

www.cambridge.org
Information on this title: www.cambridge.org/9780521358583

First published 1988
Reprinted 1992

A catalogue record for this publication is available from the British Library

Library of Congress Cataloguing in Publication data
O'Dell, T. H. (Thomas Henry)
Electronic circuit design: art and practice / T.H. O'Dell.
172 pp. cm.- - (Electronics texts for engineers and scientists)
Includes bibliographies and indexes.
ISBN 0-521-35329-7. ISBN 0-521-35858-2 (pbk.)
1. Electronic circuit design. I. Title. II. Series.
TK 7867.028 1988
621.3815'3 - - dc 19 87-32903 CIP

ISBN 978-0-521-35329-8 Hardback
ISBN 978-0-521-35858-3 Paperback

Contents

Preface

This book has grown out of the experience I gained while involved with a laboratory course in electronic circuit design for the final year undergraduates in the Department of Electrical Engineering, Imperial College, University of London. The book is intended to provide a selection of ideas for such a course, more material being described than could be fitted into a single option during one academic year, and much being suitable for the earlier years of a degree course.

The book is also intended for a far wider range of readers who have access to a simple electronics laboratory: those undergraduates, graduates, research assistants, and technicians in all fields of the hard and soft sciences who, as Horowitz and Hill put it in the Preface to their *Art of Electronics*, 'suddenly find themselves hampered by their inability to "do electronics"'.

The point which is emphasised here is that electronic circuit design can only be learnt by doing. Theoretical knowledge is, of course, essential and this book must be used with a good foundation text for the fundamental principles of electronics. Horowitz and Hill is an excellent choice for this, and the book by Gray and Meyer will cover any more advanced theory which is needed. Details of these two books will be found in note 2 of Chapter 3 here.

There is something more to electronic circuit design, however, than a good theoretical foundation coupled with a considerable amount of experience in the laboratory. This something more is the question of design itself: where do *new* circuit ideas come from? This problem is discussed in the first chapter 'What is design?', and the discussion is maintained throughout the following eight chapters, which deal with high and low frequency small signal circuits, opto-electronic circuits, digital circuits, oscillators, translinear circuits, and power amplifiers. In each chapter there are one or more experimental circuits for the reader to construct and make measurements on, a total of thirteen small project exercises in all. This practical work of construction, followed by the observation of waveforms, voltage and current levels, non-linearities, thermal effects, and so on, is the way to understand how new circuit ideas develop and why electronic circuits take the shape that they do. Then, readers will find themselves inventing new circuits of their own; new, perhaps, only to them, but that is no problem. To discover new techniques for yourself in the only valid way to keep up to date.

The final chapter 'Theory and practice' draws some brief conclusions on the fundamental problem of design, in the light of the growing literature about it and in the light of the circuits which have been dealt with in the book.

In writing this book I have made use of many ideas which have come out of discussions with colleagues, or during project and laboratory work with students. I should like to thank them all. Thanks are also due to Dr B. A. Unvala, of Imperial College, and to Mr Francis Saba, who both read the text and suggested many improvements.

T. H. O'Dell
London, January 1987

1

What is design?

1.1 Design and synthesis

As the word 'design' is used in electronic engineering rather freely it seems essential, for a beginning to our discussion of electronic circuit design, to be clear what is meant. The production of a piece of electronic hardware, starting of course with a clear idea of what that hardware is supposed to do, really involves two stages before the work of manufacture or construction can begin. At first, we must outline how we are going to solve the problem. We need a preliminary sketch of the circuit. Secondly, we need to calculate the exact component values, or values of the parameters which enable us to construct or process the circuit. Here, we shall call these two steps design and synthesis. It is not, of course, possible to separate these two steps. They interact with one another and the complete engineering process involves a continuous shift from design to synthesis and then back to design. This is what 'computer aided design' should mean.

For electronic circuits, the design step is primarily a choice of what we shall call circuit shape. The word 'shape' has been chosen with the German word for shape, *die Gestalt*, in mind because this has now come into the English language [1] to mean 'an organised whole in which each individual part affects every other, the whole being more than the sum of its parts'. Such an attribution is meant here by circuit shape, and this should become clear below when we look at some examples. Even more is involved, however, because it is not just the individual parts of a circuit, and the way in which these are connected up, which are important. As Baxandall [2] has remarked, the way a circuit is actually drawn has a very strong influence upon the way one thinks about how it works.

1.2 Choosing a circuit shape

Fig. 1.1 shows an electronic circuit, a relaxation oscillator, a circuit shape which will have been made repeatedly familiar to electrical engineering students over the past sixty years. This is an historical fact because the circuit, as a shape, probably first appeared as Fig. 62 on page 158 of a 'secret' document dated April 1918, complete with all the differential equations describing its operation [3]. At that time the circuit used two triodes and worked extremely well with these devices because, in the circuit shape of Fig. 1.1, nearly the full supply voltage, V_+, is available to reverse bias the grid with respect to the cathode when either triode has to be turned off.

Using transistors, the circuit shape shown in Fig. 1.1 is not a good choice for a design. The reason is that, in contrast to a triode, a bipolar transistor can have no reverse bias across its emitter junction until it *has* been turned off. Furthermore, while the positive supply voltage, V_+ in Fig. 1.1, may be only about 15 V, this is quite enough to break down the emitter junction of a silicon planar transistor. Fig. 1.1 may well have worked when it was first tried out with germanium transistors, but it is surprising to find that it is still presented as a worthwhile design today [4].

Consider an alternative circuit shape for a multivibrator: the circuit shown in Fig. 1.2. Like Fig. 1.1, this is a two stage amplifier with its output fed back to its input. The first stage, Q_1, is in grounded base connection, and the second stage, Q_2, is in grounded collector connection, as far as the oscillator loop is concerned. Q_2 is also used as an output buffer amplifier, which is a good idea compared to simply omitting R_4 and taking the output from the collector of Q_1.

The advance in technique, as far as making effective use of the bipolar transistor is concerned, in going from Fig. 1.1 to Fig. 1.2, cannot be overemphasized [5]. In Fig. 1.2 use is made of the small voltage across the

Fig. 1.1.

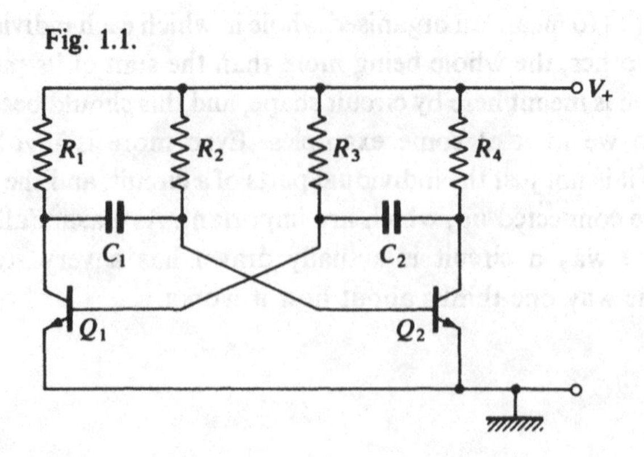

emitter junctions of Q_1 and Q_2, when these devices are conducting, to enable us to define the operating currents of Q_1 and Q_2 by means of resistor values. The transistors do not saturate in this circuit; it was the fact that Q_1 and Q_2 were allowed to saturate in Fig. 1.1 which defined the operating currents in that circuit. Full details of Fig. 1.2 are given in the Appendix because this is an interesting circuit to experiment with.

Let us look at a third possible shape for a multivibrator or relaxation oscillator circuit. This is shown in Fig. 1.3, and is the kind of circuit shape which might be found in a silicon integrated circuit [6].

Often described as an emitter coupled multivibrator, Fig. 1.3 is, in fact, more complicated than Fig. 1.2 because it includes some of the ideas shown in Fig. 1.1. We do not discuss the way these circuits work here, they are all well known. We are only proposing different circuit shapes which all can, with different advantages and disadvantages, provide the solution should we need a relaxation oscillator, or at least provide a basis to build on for a good final design.

1.3 Method in design

Is there any method that can be followed in order to come up with new circuit ideas? When the circuits shown here as Figs. 1.1, 1.2, and 1.3 were first drawn, they were without question quite new conceptions. How was it done? How can we continue to come up with good new circuit ideas?

There have been attempts in the past to find a method in the design of electronic circuits, and by design we again mean the invention of the shape of a circuit, the step which must take place before any problems concerned

Fig. 1.2.

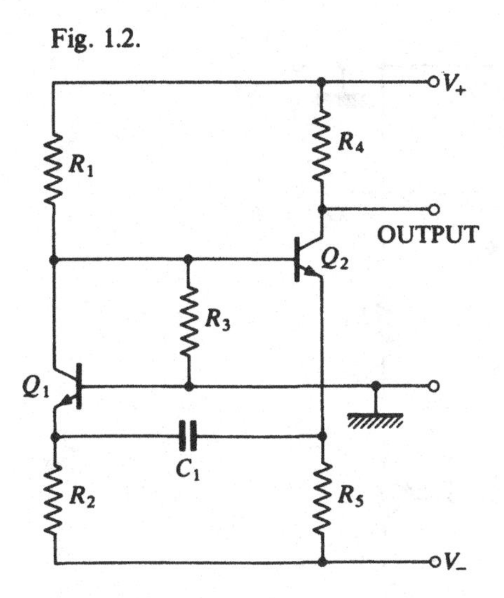

with component values, that is with synthesis, come up. The use of signal flow graphs and network topology was at one time considered a step in the right direction, but both these techniques, while making important advances which helped with analysis and synthesis [7], contributed nothing to our understanding of the design process. It is, perhaps, interesting to look at one of the texts which introduced signal flow graphs and network topology to the main stream of electronic engineers, and to quote from the preface of the first edition, dated 1960:

> The laws of physics provide basic models of charge motion, in terms of which we explain the observed terminal behaviour of elementary components and devices . . . From physical models and observed terminal characteristics, we make circuit models that lead to simple circuits capable of performing basic operations. . . . These operations . . . serve as elementary building blocks or models with which we can construct models of more general systems [8].

This is a very good example of the feeling that technology is just a matter of applying science, a feeling which was very much the accepted one in the middle of this century. Today, we know little more about the problem except that it is far more complicated. Technology is certainly not just applied science, but, as Mayr has written [9], 'the problem of the science–technology relationship . . . has eluded all who have tried to grasp it', and he goes on to list the great variety of people who have tried. In their very interesting book, *The Sources of Invention*, Jewkes, Sawers and Stillerman consider an even wider field and write 'The interactions between science, technology and economic growth are much more complicated than was

Fig. 1.3.

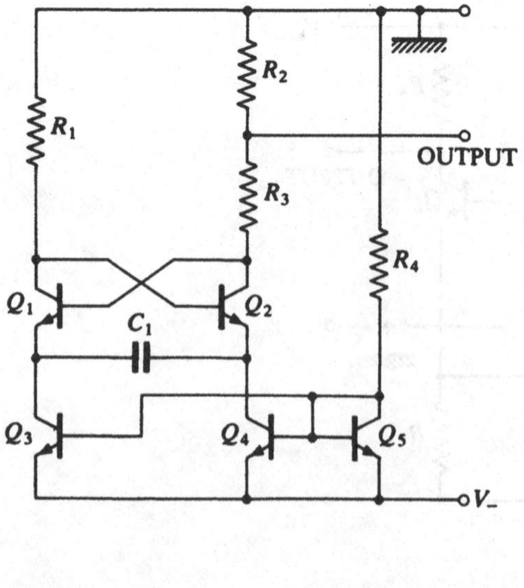

ever imagined by those who have dominated opinion and influenced public policy upon these matters in recent years' [10].

So we are left with our original difficulty when we ask where new circuit shapes, new circuit ideas, actually come from. This difficulty might be answered with one word: '*eureka*', 'I have found it'. However, we must be careful not to then make the assumption that there is one particular person responsible for any new circuit idea. The truth is far more likely to be that a new circuit shape has appeared in a number of different places, all at about the same time, simply because so many workers in the field have been thinking along the same lines. T. S. Kuhn has written about the same thing happening with scientific innovation [11], as opposed to the technological innovation we are concerned with here.

In the following chapters we shall be looking at some of the interesting circuit shapes that various people have come up with, often trying to trace the origin of these new circuits by looking at the kind of circuit shapes which were used earlier. As far as possible, circuits have been chosen so that the reader can construct them in a laboratory which has only the simplest facilities. This has been done because an experimental point of view seems to be essential for making progress in circuit design, for beginning to find new circuit shapes even if these are new only to the experimentalist. From the point of view which is taken in this book, a circuit designer comes from the laboratory, not from the lecture theatre.

1.4 Working with circuit shapes

While there appears to be no clear method behind the invention of a completely new circuit, it does seem possible, in some cases, to find a pattern in the development of a new idea.

This pattern can take its form from simply adding simpler circuits together. We have an example here already in Figs. 1.1, 1.2, and 1.3: the idea of two stages of common emitter amplification, Fig. 1.1, and the idea of emitter coupling via a capacitor, Fig. 1.2, both appear in Fig. 1.3. It might be said that the circuit shape of Fig. 1.3 comes from embedding Fig. 1.2 into Fig. 1.1. This idea of embedding one circuit into another is explored further in Chapter 3. The simpler idea of just adding circuits, one after another, is looked at first, in Chapter 2, and also in Chapter 4 where we look at the kind of new circuits people have come up with by simply adding operational amplifier circuits together.

The idea of working with circuit shapes may be taken even further when integrated circuits are considered. This comes about because the idea of adding circuits together, or embedding one circuit into another, takes on a far deeper meaning when we deal with silicon, n^+, n, i, p and p^+, and not just with the nodes, or input and output terminals, of circuits which, while they may not be discrete component circuits, are still thought of in a discrete way. Some very interesting examples of this kind of work with circuit shapes come up in Chapter 6 when we look at the invention of integrated injection logic and substrate fed logic.

1.5 Construction and manufacture

The way a circuit is going to be made has a very great effect on the kind of circuit shape we may finally end up with. Of our examples so far, Figs. 1.1 and 1.2 are classical discrete component circuits, while Fig. 1.3 would be much better made as an integrated circuit, or at least by using an array of transistors [12], because Q_3, Q_4 and Q_5 need to be in close thermal contact.

It is from this aspect, constructional and manufacturing technique, that development and technical change in electronics are most apparent. However, while references given to particular devices, and even to kinds of device, will always become rapidly out of date, the central problems of electronic circuit design have been surprisingly constant for a very long time. Unless we have just entered a 'post historical' era [13], the electronic hardware of today should look as old fashioned, to the electronic engineer of fifty years from now, as the electronic hardware of World War II looks to us when we see it in our museums today. Whatever devices or constructional techniques we may use in the future, we shall still be building on the theory which has its foundations in the past, and we shall probably still be trying to understand how people come up with such elegant and simple ideas, ideas which, with hindsight, make us feel that the whole problem is quite easy and must be obvious.

Notes

1 **Oxford Concise Dictionary**, Oxford, 1970, p. 513.

2 P.J. Baxandall, *Radio and Electronic Eng.*, **29**, 229–46, 1965.

3 This document is the remarkable 'Notice sur les lampes – valves à 3 électrodes et leurs applications', Ministère de la Guerre, Establissement Central du Matérial de la Radiotélégraphie Militaire, Avril 1918. Copies of this are held in the Centre de Documentation d'Histoire des Techniques in Paris, ref. DOC 1359, and in the IEE Archives in London, ref. NAEST 46/9.

4 In an issue of *IEEE Spectrum*, No. 11, 1984, which is devoted to contemporary education, the circuit shown in Fig. 1.1 appears on page 48 as a photograph which has been taken from the VDU display of a 'Circuit Analysis Program' designed for students.

5 The circuit shown in Fig. 1.2 may have been first described by R.C. Bowes,

Proc. IEE, **106B**, *Suppl.* 15–18, 793–800, May 1959. Bowes gives some comments on the possible origin of this circuit.

6 The circuit shown in Fig. 1.3 is a very much simplified version of the one used in the 560 series of phase locked loops introduced in 1970 by Signetics: *Linear Integrated Circuits*, Signetics Int. Corp., London, 1972, p.260.

7 F.D. Waldhauer, *Bell Syst. Tech. J.*, **56**, 1337–86, 1977.

8 S.J. Mason and H. Zimmerman, *Electronic Circuits, Systems and Signals*, John Wiley, New York, 1960. This book covers network topology in Ch. 3 and signal flow graphs in Chs. 4 and 6.

9 O. Mayr, *Tech. and Culture*, **17**, 663–73, 1976.

10 J. Jewkes, D. Sawers and R. Stillerman, *The Sources of Invention*, Macmillan, London, 1969, 2nd edition, p.226.

11 T.S. Kuhn, *Hist. Studies Phy. Sciences*, **14**, 231–52, 1984, particularly pp.250–1.

12 A variety of these are available with up to seven devices (CA3081), using bipolar devices of one or the other polarity, or even mixed (CA3096). There are also arrays of MOS transistors (CA3600).

13 In his very interesting paper, 'Historians and Modern Technology', *Tech. and Culture*, **15**, 161–93, 1974, Reinhard Rürup warns against a 'technological mode of thinking – a way of thinking according to which man's hope and future lie in the end of history'. Just such a mode of thinking is to be found in Dennis Gabor's *Inventing the Future*, Secker and Warburg, 1963: Chapter 10 is subtitled 'History must have a stop'.

2

High frequency band-pass amplifiers

2.1 Tuned amplifiers

We begin our course on electronic circuit design with what must be the simplest and most classical of all electronic circuits: the tuned amplifier. This is the kind of circuit which is found at the front end of a good communications receiver [1] to give a well defined power gain over a well defined range of frequencies, and also to have the lowest possible noise figure. Amplifiers of this kind are also found in many of the most modern scientific instruments, such as magnetometers using superconducting detectors or nuclear magnetic resonance systems which are used in analytical chemistry and in medical diagnostics.

In the past, tuned amplifiers of the kind we shall discuss were also found as the intermediate frequency amplifiers of radio, radar, and television receivers, but modern designs can be much simpler because of the very cheap and effective filters which have become available [2]. These make it possible to use an amplifier with a rather crude band-pass characteristic, and then correct this with a good filter. This approach is only possible when high noise figure and low efficiency can be tolerated.

2.2 The problem of regeneration

The main problem that has always plagued the design of high frequency band-pass, or tuned, amplifiers is that unwanted feedback from output to input may cause a radical departure from the expected performance, and can even cause complete instability of the amplifier so that it just oscillates close to the frequency where it should amplify. Let us see how this problem relates back to the problem of circuit shape, which we discussed in the first chapter.

In Fig. 2.1(*a*) we show a grounded emitter configuration of a bipolar transistor. This is connected as an amplifier with input voltage, V_{in}, output voltage, V_{out}, and a load, Z_L, which in a tuned amplifier will be some kind of resonant circuit. Undesired feedback from output to input can obviously occur in at least two ways in Fig. 2.1(*a*): either via the capacitance which must exist between the collector and the base, or because of the unavoidable impedance of the emitter and its connection to ground. The latter is fairly easy to get rid of but capacitance between collector and base, the capacitance of the collector junction, is just an essential part of the transistor. We shall go into the theory of why this capacitance causes problems later in this chapter. For the moment, we have a design problem, not a problem of synthesis, and we begin to look for a new circuit shape idea which might remove the problem of the collector junction capacitance producing unwanted feedback.

The mainstream traditions of electronic engineering make the majority of people feel that the grounded emitter configuration, Fig. 2.1(*a*), is the natural way to use the bipolar transistor as an amplifier [3]. This was certainly the feeling with the earlier amplifying devices which used hot emitters: the most sensible thing to do with the heater–cathode part was to ground it! However, it is rather amusing to remember that the first transistor had an equally embarrassing *base*, that was why it was called a 'base', and the most natural thing to do with the transistor was to use it in the grounded base configuration shown in Fig. 2.1(*b*) [4].

Fig. 2.1(*b*) shows a very simple change, from Fig. 2.1(*a*), in what we are calling circuit shape. Here we have the grounded base configuration where the signal to be amplified is fed to the emitter and the output, as in Fig. 2.1(*a*), appears across the load impedance, Z_L, which is connected to the collector. The grounded base now lies in between the output and the input

Fig. 2.1.

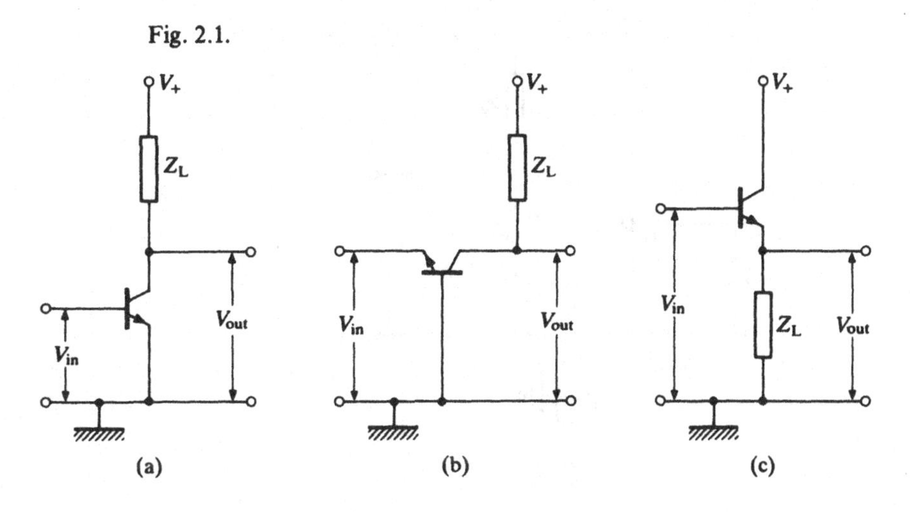

so that it should be possible to make the unwanted feedback capacitance very small indeed.

Fig. 2.1(*c*) shows the third possible change we can make in circuit shape just by shifting around the two components, a transistor and a load impedance. This circuit has a voltage gain slightly less than unity, and it is not phase inverting, so the input and output terminals are always at very nearly the same potential and unwanted capacitive feedback from output to input is not normally a problem.

2.3 Adding circuits together

Let us continue with this design, that is, choosing a circuit shape to use for our tuned amplifier, simply drawing circuit ideas but now working with the three circuits shown in Fig. 2.1 and adding these together in different ways. We are looking for a circuit which should have minimum feedback from output to input.

Fig. 2.2 shows the result of using Fig. 2.1(*c*) to drive Fig. 2.1(*b*). The load impedance of the first transistor is now the input impedance of the second, so that anything else which we have to connect to this common connection must have a very high impedance, and we show a constant current generator sinking a bias current, I_B, down to a negative rail. This circuit is a really good design. The feedback capacitance between output and input, that is, between the collector of Q_2 and the base of Q_1, can be made very small when the circuit is produced as a silicon integrated circuit, as it has been for many years now as the RCA CA3028A/B. We shall give a detailed design using this device at the end of this chapter. The circuit shape,

Fig. 2.2.

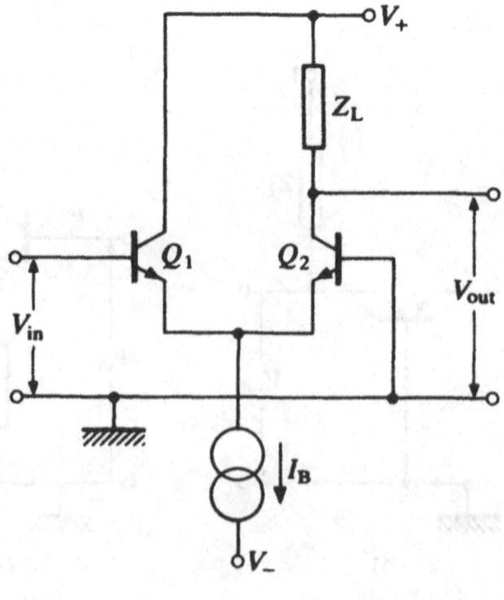

however, has a long tradition behind it [5] and, when it has two balanced loads, is often called a 'long tailed pair': a circuit which will play a central role in Chapter 3.

Fig. 2.3 shows the result of using Fig. 2.1(a) to drive Fig. 2.1(b). We have to ground the base of Q_2 through a capacitor, C, because this base has to be fixed at some d.c. level between V_+ and ground. The circuit is well known as the 'cascode' [6] and is another excellent circuit, particularly when it is realised as a silicon integrated circuit using MOS technology. The device then becomes a dual gate MOST (metal oxide semiconductor transistor) because the source of Q_2 merges into the drain of Q_1 and this part of the circuit need not be connected to the outside world. Dual gate MOST devices are widely used as high frequency tuned amplifiers up to 500 MHz.

2.4 An experimental circuit

Now let us use the circuit shape of Fig. 2.3 as a design idea for a real experimental circuit. Before making any detailed analysis of what we might expect to happen let us just try something out, because this is the way to learn about things. It is, of course, absolutely essential to go back and do the necessary theoretical work but, in many cases, it is not possible to ask the right theoretical questions before some practical experimental results have been obtained.

The circuit shown in Fig. 2.3 uses two bipolar transistors. For reasons given at the end of the last section, we shall choose to use a dual gate MOST instead. We shall also choose to build an amplifier that can be inserted into a 50 Ω or 75 Ω transmission line, working at a modest frequency of 10 MHz, so that experimental work will be possible in the most ordinary laboratory using a simple signal generator for a source and an oscilloscope

Fig. 2.3.

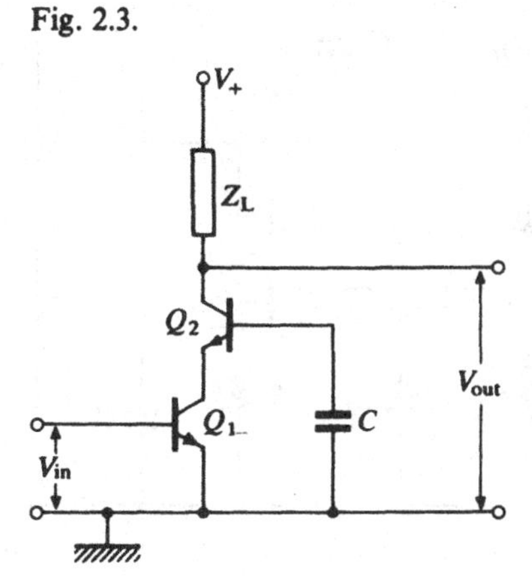

or r.f. voltmeter, with a proper termination, to measure the amplifier output.

The RCA 40841 MOST will be suitable for such an application and the data sheet shows that this device should be operated at a drain to source voltage of +15 V, a gate 1 to source voltage of −0.5 V and a gate 2 to source voltage of +4 V. The drain current should then be close to 5 mA and the forward transconductance, g_m, over 10 mS [7].

The complete circuit for the proposed experiment is shown in Fig. 2.4. The bias conditions described above, for $V_+ = 15$ V, will be obtained if we make $R_1 = 10$ k, $R_2 = 4.7$ k and $R_3 = 100\,\Omega$. C_3, C_4, and C_5 are simply decoupling capacitors. To operate near 10 MHz, let us first choose $C_1 = C_2 = 50$ pF and then we find that L_1 and L_2 should be forty turns, close wound in a single layer on a 6 mm former to give a winding 6 mm long. This means a wire diameter of about 150 μm. The coils shown in Fig. 2.4 which couple the input and output cables would need to be about five turns, wound close to the grounded ends of L_1 and L_2. L_1 and L_2 have adjustable cores to give a small variation in inductance.

More detail on how this circuit might actually be built up is given in the Appendix. Every experimental circuit of this kind which can be made, using slightly different layouts or constructional techniques, will have its own individual peculiarities, and this is just what makes electronics so interesting.

Fig. 2.4.

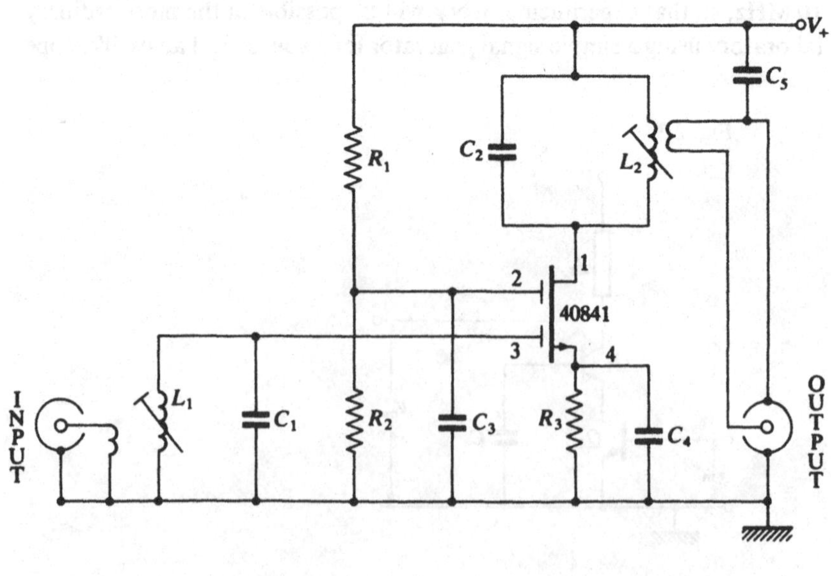

2.5 Performance of the experimental circuit

Let us suppose that both resonant circuits shown in Fig. 2.4 have a Q factor of 100. This is mainly determined by the source impedance and load impedance presented by the input and output cables. Let us further suppose that, by some miracle of ingenuity, we have been able to tune the two resonant circuits to exactly the same frequency, 10 MHz. What would we expect to find when we connected the circuit to its +15 V supply and checked its performance?

Curve (1) in Fig. 2.5 shows what we would expect to observe as a gain–frequency characteristic, assuming that the dual gate MOST had negligible feedback capacitance from drain to gate 1. The curve shows a perfectly symmetrical pass-band, centred on 10 MHz, with a 3 db bandwidth of 64 kHz, as we would expect from two circuits, both of $Q = 100$, connected in cascade. The maximum gain is normalised to unity in Fig. 2.5 for convenience.

Now, still maintaining that we were somehow able to set both resonant circuits to 10 MHz beforehand, let us move one step closer to the real world and admit that we shall have to tolerate at least 0.02 pF feedback capacitance with the 40841 MOST, because this is the figure given in the

Fig. 2.5.

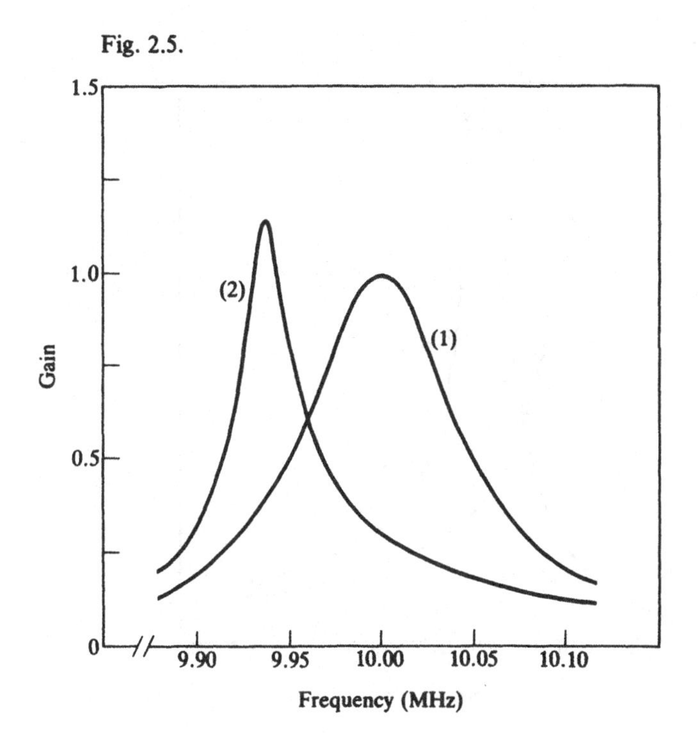

data sheet [7] for the device on its own. Fig. 2.5 shows what we would then observe to be the pass-band characteristic as curve (2). The maximum gain is increased a little, the bandwidth is very much smaller and the pass-band is centred on a frequency well below 10 MHz. What has happened?

The commonsense answer to that question is found by going back to our initial assumption that both tuned circuits are really tuned to 10 MHz. It would, of course, be quite impossible to arrange this before any measurements were made. We could only do this to a fair degree of accuracy by making the two circuits as identical as possible and then trying to compensate for the slightly different additional capacitance that is added by the transistor and circuit layout.

It follows that our first reaction to the experimental result shown as curve (2) in Fig. 2.5 would be that one, or both, of our tuned circuits was, in fact, tuned to a frequency slightly below 10 MHz. All we need to do is reduce L_1 or L_2 slightly, using the adjustable cores, and all will be well.

If we did take this action, on this particular circuit, the result might well surprise us because, although the maximum gain would move towards 10 MHz, it would also begin to grow very rapidly and the bandwidth would get very narrow indeed. Fig. 2.6 shows what we would observe.

Fig. 2.6.

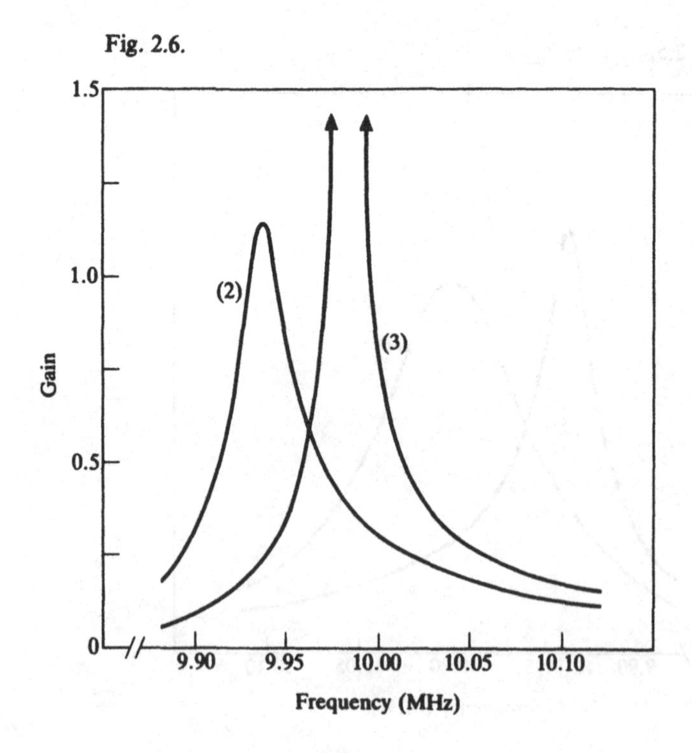

In Fig. 2.6, curve (2) is the same situation as curve (2) in Fig. 2.5: both tuned circuits are at exactly 10 MHz. Curve (3) shows the result of adjusting L_1, in Fig. 2.4, to a slightly smaller value so that the input tuned circuit is now resonant at 10.13 MHz. The maximum gain is now well over 9, right off the scale of our diagram, and the bandwidth is very small indeed. Further adjustment of L_1, close to this new setting, will cause the circuit to oscillate. In other words, the gain will become infinite [8].

2.6 Thinking with the help of vectors

We must now find out why the very small feedback capacitance from the drain to gate 1 of the device shown in Fig. 2.4 has such a dramatic effect on the performance, and we can do this very easily by drawing a vector diagram. However, this vector diagram will not be given in this text. Readers are asked to sketch it for themselves because the point becomes clear only as the diagram develops.

Let us take as our reference vector the voltage appearing across C_1. Because C_4 is very large, this voltage is V_{g_1-s} so let us draw a short vector, V_{g_1-s}, horizontally, $\rightarrow$. When C_2L_2 is exactly resonant at the input frequency we may argue that the drain load impedance, Z_L, is real and the voltage at the drain, that is, V_{d-s}. is $-g_m Z_L V_{g_1-s}$: a long horizontal vector opposed to V_{g_1-s}.

Now V_{d-s} is by far the largest component of voltage across the small feedback capacitor C_{rs}. It follows that a current $-j\omega C_{rs} g_m Z_L V_{g_1-s}$ will be fed back to gate 1 and this is a short vector vertically downwards, $\downarrow$. With both resonant circuits perfectly in tune with the input signal, this small current can only produce a voltage at the drain which has a +j signature, that is $\uparrow$, and repeating the argument above we can see a second order feedback which will reduce our effective V_{g_1-s} and thus explain the drop in gain at 10 MHz as we move from curve (1) to curve (2) in Fig. 2.5.

Why should the gain increase, as shown by curve (2) in Fig. 2.5, as we reduce the input frequency? The answer may be seen at once when we remember that the impedance Z_L has a real part, R, in parallel with an imaginary part, $j\omega L/[1-(\omega/\omega_0)^2]$, where ω_0 is the resonant frequency, $1/\sqrt{LC}$. At a frequency below resonance, $\omega < \omega_0$, both resonant circuits look inductive and the voltage across them will lead the current by a small angle. Let us go back and change our vector diagram so that the voltage at the drain, $-g_m Z_L V_{g_1-s}$ is not horizontal any more but has a small component downwards. The current fed back will still be at 90° to this voltage, since C_{rs} is a pure capacitance, so it will now have a component in phase with V_{g_1-s}, that is positive feedback. The voltage produced across L_1C_1 by this current will have an even larger component in phase with

V_{g_1-s} because this input circuit is also looking inductive and the voltage across it will lead the current fed to it.

This qualitative argument, made by sketching with vectors, shows us why we have problems with the simple circuit of Fig. 2.4. It also explains why things get dramatically worse, as shown in Fig. 2.6, when we make the input tuned circuit look even more inductive by increasing its resonant frequency. But this argument does not help at all in telling us what we must do to make this circuit usable.

2.7 What is a usable circuit?

The concept of usability in electronics is rather subjective and has been subject to historical change [9]. The reasons for this are due to changing manufacturing techniques and how electronics is used, but in recent times a good definition of usability is that a circuit can be manufactured. Clearly, a circuit cannot be said to be usable, from this point of view, if it behaves in the way our present experimental circuit is behaving, as shown by Fig. 2.6. If changing the resonant frequency of L_1C_1 from 10.00 MHz to 10.13 MHz, only just over 1%, changes the gain and bandwidth so dramatically, the circuit cannot possibly be manufactured and adjusted to give some well defined performance which can be expected to be fairly constant in time.

Having chosen the particular circuit shape shown in Fig. 2.4 for our design, we now need some simple relationship, if one exists, to tell us how to choose the component values. For example, we know that L_1 and C_1 must be chosen so that $\omega_0/2\pi$ is close to 10 MHz but what of the ratio L_1/C_1? Remember that our original choice of $C_1 = C_2 = 50$ pF was quite arbitrary.

We might consider treating this problem numerically [10] but, as argued in Chapter 1, numerical calculations belong to circuit synthesis. For design an algebraic standpoint must be taken.

2.8 Algebra and design

To understand the constraints which must be placed upon our choice of components in the circuit shown in Fig. 2.4, we need to consider its essential features at high frequency and then ask what happens as we vary all the component values. This can be done by using the very simple model shown in Fig. 2.7, in which the input tuned circuit is the two terminal network A–B, the transistor is the two port network y_{ij}, and the output tuned circuit is the two terminal network C–D. Note that the circuit Q is introduced by the LCR representation in which $R = Q\sqrt{L/C}$.

The amplifier circuit of Fig. 2.4 is represented by Fig. 2.7, as far as its high frequency, small signal, behaviour is concerned, when we connect A, B to 1,

2 and 3, 4 to C, D. To understand the behaviour shown in Figs. 2.5 and 2.6, however, we only need to disconnect the input tuned circuit and consider the input admittance, Y_{in}, looking into 1, 2 when the output tuned circuit is connected [11].

To do this we take

$$i_1 = y_{11}v_1 + y_{12}v_2 \tag{2.1}$$

$$i_2 = y_{21}v_1 + y_{22}v_2 \tag{2.2}$$

for the device, and

$$i_2 = -Y_L v_2 \tag{2.3}$$

for the output tuned circuit, where

$$Y_L = (Q_2\sqrt{L_2/C_2})^{-1} + j(\omega C_2 + 1/\omega L_2) \tag{2.4}$$

Now (2.2) and (2.3) give

$$v_2 = -[y_{21}/(y_{22} + Y_L)]v_1 \tag{2.5}$$

and (2.5) with (2.1) gives

$$Y_{in} = i_1/v_1 = y_{11} - y_{12}y_{21}/(y_{22} + Y_L) \tag{2.6}$$

In general all the admittances on the right-hand side of equation (2.6) will vary with frequency. We have to use equation (2.6) to calculate how Y_{in} varies with frequency and if, over any frequency range, the real part of Y_{in} becomes more negative than $(Q_1\sqrt{L_1/C_1})^{-1}$ then the amplifier will oscillate when the input tuned circuit is connected, simply because the total damping of this input circuit is now positive.

Now this calculation can become quite complicated but with an MOST, at not too high a frequency, we can accurately represent the device as having a real transconductance,

$$y_{21} = g_m \tag{2.7}$$

and a pure feedback capacitance,

$$y_{12} = -j\omega C_{rs} \tag{2.8}$$

Fig. 2.7.

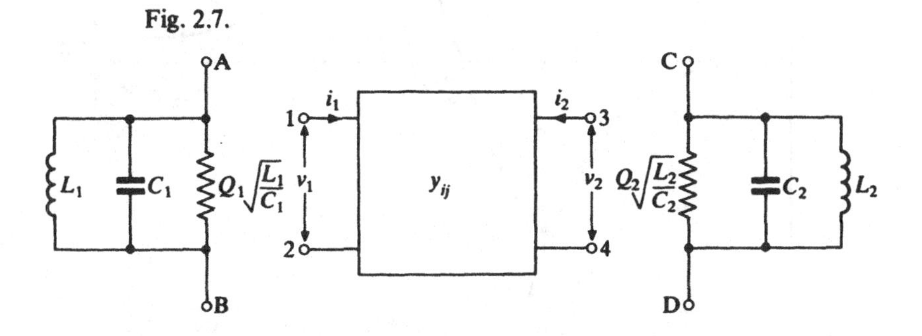

and treat y_{11} and y_{22} as zero because both can be included as small corrections to C_1 and C_2. Note that the signs in equations (2.7) and (2.8) are determined by the directions of i_1 and i_2 in Fig. 2.7.

Using equations (2.7), (2.8), and (2.4) in (2.6), and writing $\omega_2 = 1/\sqrt{L_2C_2}$, we obtain

$$Y_{in} = \frac{-(g_m C_{rs}/C_2)(\omega/\omega_2)}{(\omega_2/\omega - \omega/\omega_2) + j/Q_2} \tag{2.9}$$

and this has a real part,

$$\text{Re}\{Y_{in}\} = \frac{(g_m C_{rs}/C_2)(\omega/\omega_2)(\omega/\omega_2 - \omega_2/\omega)}{(\omega_2/\omega - \omega/\omega_2)^2 + Q_2^{-2}} \tag{2.10}$$

Now it is here that the importance of an algebraic approach to this problem becomes clear. We need a sketch of the function given by equation (2.10): a function of ω/ω_2, the working frequency normalised to the resonant frequency of the output circuit. This sketch is shown as Fig. 2.8.

By differentiating equation (2.10) with respect to ω/ω_2, we find that the function has a maximum value when $\omega/\omega_2 = (1 + 1/2Q_2)$ and a minimum value when $\omega/\omega_2 = (1 - 1/2Q_2)$. As shown in Fig. 2.8, the real part of Y_{in} does go negative, and the most negative value is $-g_m C_{rs} Q_2/2C_2$. It follows that the circuit will oscillate unless we ensure that

$$g_m C_{rs} Q_2/2C_2 < (Q_1\sqrt{L_1/C_1})^{-1} \tag{2.11}$$

Inequality (2.11) takes on a particularly simple form when we consider a particularly simple circuit, like Fig. 2.4, in which $C_1 = C_2 = C$ and $Q_1 =$

Fig. 2.8.

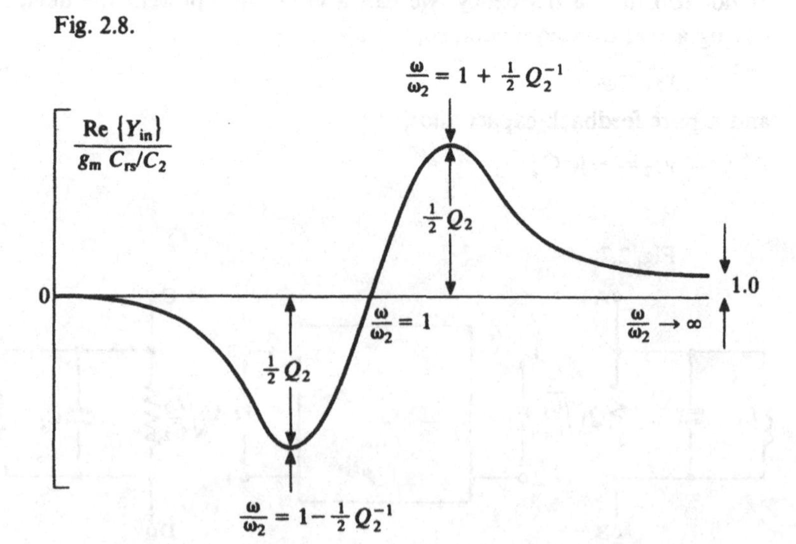

$Q_2 = Q$. Let us then write $\omega_0 = 1/\sqrt{LC}$ for the operating frequency and inequality (2.11) becomes

$$g_m C_{rs} Q^2 / 2\omega_0 C^2 < 1 \qquad (2.12)$$

2.9 Making the experimental circuit stable

Let us now use inequality (2.12) to see just how bad our original choice of component values for Fig. 2.4 was. Our choice was $C = 50$ pF and $Q = 100$. With C_{rs} at 0.02 pF, $g_m = 10$ mS and operating at 10 MHz, the left-hand side of inequality (2.12) comes out at $20/\pi$, much greater than unity. It is now obvious why we ran into the problems illustrated in Figs. 2.5 and 2.6.

In order to get a stable amplifier we look at inequality (2.12) and realise that, for example, we should increase the value of C and reduce the value of Q. This would be done by reducing L and increasing the coupling between the main inductors and the small coils, shown in Fig. 2.4, which couple the input and output cables.

An example would be to increase C to 100 pF and reduce Q to 50. With the other values unchanged, the left hand side of inequality (2.12) is well below unity at $4/8\pi$. The performance of the amplifier is shown in Fig. 2.9,

Fig. 2.9.

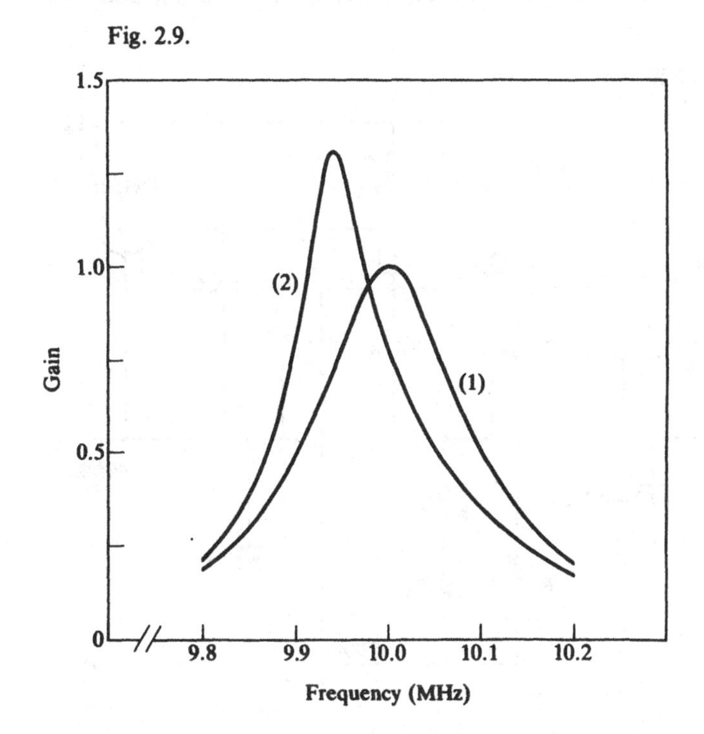

where curve (1) shows the performance that would be obtained if there were no feedback capacitance, $C_{rs} = 0$, and curve (2) shows the performance of the real amplifier when both tuned circuits are adjusted to 10 MHz. We find that curve (2) does, in fact, give the maximum gain under these conditions and to tune the amplifier to the desired 10 MHz we simply have to adjust both tuned circuits to a slightly higher frequency. As we do this, the gain never increases, unlike the behaviour shown in Fig. 2.6.

Real synthesis, or computer aided design work, could now begin on this amplifier. The new choice of components, which led to the performance shown in Fig. 2.9, is still a pretty arbitrary one and the circuit may be too sensitive to component value changes and layout and, anyway, is unlikely to have exactly the bandwidth, gain, and noise figure required.

2.10 An experimental circuit using bipolar transistors

To conclude this chapter, let us look at another 10MHz tuned amplifier, this time using the circuit shape of Fig. 2.2 and using bipolar transistors.

The circuit is shown in Fig. 2.10 and uses the CA3028A, a very simple integrated circuit containing three transistors and three resistors. The CA3028A, as it is connected in Fig. 2.10, is identical to Fig. 2.2 with Q_3, R_1, R_2, and R_3 taking on the task of the bias sink, I_B. All we need to add are the

Fig. 2.10.

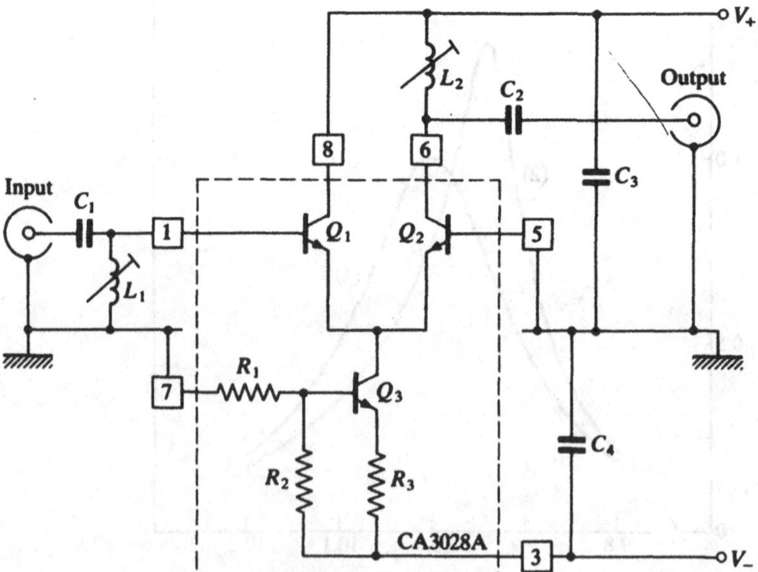

input and output tuned circuits, L_1C_1 and L_2C_2, and the power supply decoupling capacitors, C_3 and C_4. The circuit operates on ± 10 V supply lines and is a very interesting and instructive one to experiment with. Full constructional details are given in the Appendix.

Now the situation in this circuit is very different from the MOST circuit of Fig. 2.4 because the y parameters of a bipolar device are much more complicated and interesting. Let us look at the data [12] for the unwanted feedback, y_{12}. This is given explicitly as (0.01 − j0.0002) mS, for the connection shown in Fig. 2.10, at a frequency of 10.7 MHz. It is obvious that we are not dealing with a simple capacitative feedback: at 10.7 MHz y_{12} is predominantly real. The other interesting point is that −j0.0002 mS represents a capacitance of only 3×10^{-15} F, nearly a tenth of the unwanted feedback capacitance of the MOST we looked at before. How is it possible to obtain such a very small feedback capacitance?

The answer to this question is clear when we cut the top off a CA3028A and examine it under the microscope. What we see is sketched in Fig. 2.11. The output pad, 6, is in the top right-hand corner of the die and, referring to the circuit, Fig. 2.10, pad 6 has two grounded pads, 7 and 5, on either side of it. The input pad, 1, and note how as much metal as possible is removed from this pad, is similarly surrounded by r.f. grounds 8 and 3, along with the unused pads 2 and 4 which have low values of resistance connecting them to r.f. grounds.

Unwanted feedback from 6 to 1 can take place directly, via the very small capacitance which must remain and also via the isolated metallisation

Fig. 2.11.

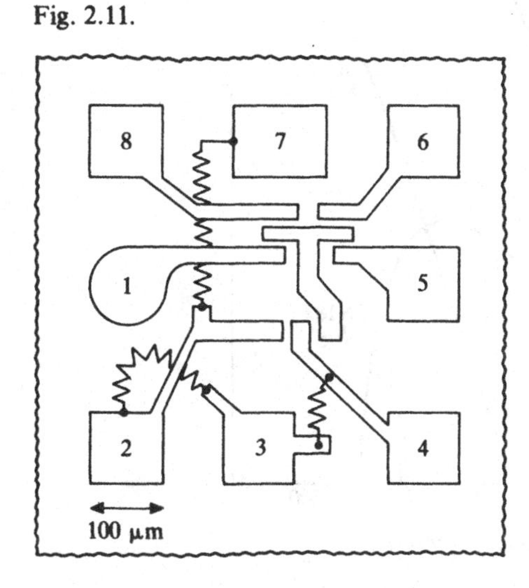

shown in the centre of the die: the two emitters of Q_1 and Q_2 and the collector of Q_3. Now feedback from 6 to this area is of the opposite phase, in its effect, compared to feedback directly from 6 to 1. It follows that a careful design of the metallisation can produce very low total feedback capacitance, at least at low frequency.

Unlike the MOST, however, things get much more complicated with a bipolar device at high frequencies. This is illustrated in Fig. 2.12 where, writing $y_{12} = g_{12} + jb_{12}$, we show how y_{12} varies over the 10 MHz to 300 MHz range for an MOST very much like the 40841 [13] and also for the CA3028A when it is used in a circuit like Fig. 2.10.

Fig. 2.12 shows that the MOST has a purely capacitive y_{12} in the 10 MHz region and, as $-b_{12}$ increases with unity slope on this log–log plot until we approach 100 MHz, we can be confident that $y_{12} = -j\omega C_{rs}$, with C_{rs} equal to 0.02 pF, is a reasonably accurate model up to 100 MHz. At frequencies above 100 MHz the picture changes completely and y_{12} becomes predominantly real.

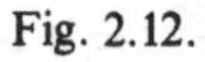

Fig. 2.12.

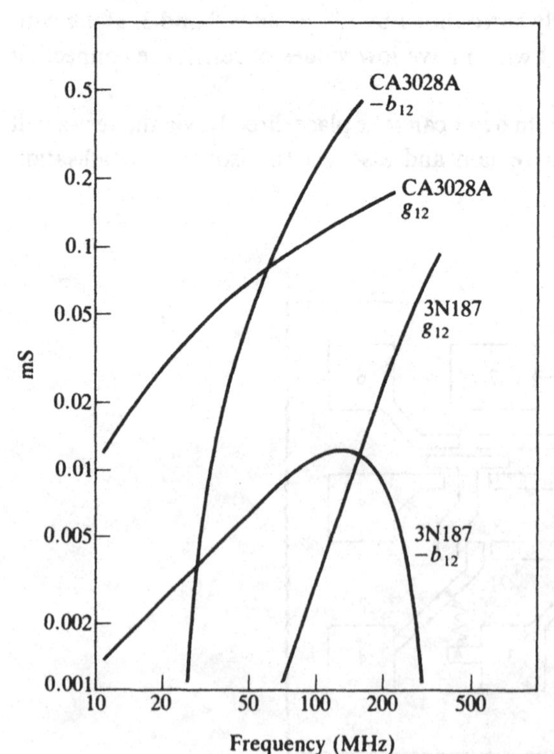

Fig. 2.12 shows that the bipolar device is quite different. At low frequencies, y_{12} is predominantly real and the imaginary part may be very small but it is increasing with frequency very much more rapidly than the susceptance of a simple capacitor. This is, of course, the lag which is developing, as the frequency increases, between the collector junction width of Q_2 and its collector voltage.

The really important point, however, is that g_{12} is positive for the CA3028A. The transconductance of the CA3028A, in the connection of Fig. 2.10, is negative, however: an increase in input voltage causes a drop in the output current. As we can represent the bipolar device by exactly the same y parameter model, Fig. 2.7, as we used for the previous circuit, we see that this means y_{21} has a negative real part at low frequency, in contrast to equation (2.7) for the MOST. The value of g_m is much larger for the CA3028A: about 40 mS [12].

This has very interesting implications. Referring to equation (2.6), we see that the product $y_{12}y_{21}$ will now have a negative real part and this will have a stabilising effect on the circuit when everything is in resonance and $(y_{22} - Y_L)$ is real. In our previous example, this kind of thing could only happen off resonance. Off resonance, the problem is very similar to the previous case with the MOST except that the curve sketched in Fig. 2.8 is mirrored in the ω axis because of the change in sign of the real part of y_{21}. Unwanted feedback with the circuit shown in Fig. 2.10 now shifts the frequency response curves to the right, instead of to the left as in Figs. 2.5 and 2.9.

The two devices discussed in this chapter have been the subject of two excellent RCA application notes [14, 15] which go into the problems which come up in the VHF range. In both the experimental circuits considered here we are dealing with very well established, or classical, electronic problems using fairly recent devices. The internal structure of these devices comes from the combination of the simple circuit shape ideas which originate from the very beginnings of electronics.

Notes

1 P. Horowitz and W. Hill, *The Art of Electronics*, Cambridge University Press, 1980, Ch. 13, p. 575.

2 The first volume in a developing series: *Wiley Series on Filters*, is *Mechanical Filters in Electronics* by R.A. Johnson, John Wiley, New York, 1983. Note also *Modern Filter Theory and Design*, edited by G.C. Temes and S.K. Mitra, John Wiley, New York, 1973, and for the higher frequencies see *Surface Wave Filters*, edited by H. Matthews, John Wiley, New York, 1977.

3 This may be because the majority feel an amplifier should have voltage gain and also have a high input impedance. The fact that power gain, coupled with good noise performance, is what is really interesting is quite an advanced idea. The use of the circuit shapes shown in Figs. 2.1(*b*) and (*c*), naturally using high vacuum devices, came quite late into electronics, a fact which is hard to explain simply on the grounds of having to wait for an indirectly heated cathode. Fig. 2.1(*c*) may be credited to A.D. Blumlein, Brit. Pat. Spec. 448421, applied for 4 September 1934, and the very sophisticated idea of combining both Figs. 2.1(*b*) and (*c*), as Fig. 2.2, is also found in one of Blumlein's patents: Fig. 3 of Brit. Pat. Spec. 482740, applied for 4 July 1936. Blumlein was not using these circuits for high frequency work however. The idea of using Fig. 2.1(*b*) on its own seems to be later: in Japan with N. Tanaka, *Nippon Elect. Comm.*, **19**, 192, January 1940, and in the UK reported in *Electronics*, **13**, 14–16 and 55–6, July 1940.

4 E. Braun and S. MacDonald, *Revolution in Miniature*, Cambridge University Press, 1978. Brattain's notes from 23 December 1947 are reproduced on p. 47 and show the first transistor amplifier circuit used.

5 Its origin in 1936 was noted above in note 3.

6 The word 'cascode' was probably coined by F.V. Hunt and R.W. Hickman in *Rev. Sci. Inst.*, **10**, 6–21, 1939. These authors, at the Cruft Laboratory, Harvard University, used it at low frequencies. The reason for the importance of the cascode, over the classical 'pentode' solution to the unwanted feedback capacitance problem in high frequency amplifiers, is given by H. Wallman *et al.*, *Proc. IRE*, **36**, 700–8, June 1948, and is a question of noise figure.

7 RCA 40841: Data Bulletin File No. 489.

8 Close to this condition for oscillation, very high power gains and very narrow bandwidths are obtainable and this was the basis of the regenerative radio receiver used in radio communications around 1914. The operator required considerable skill and experience but in later years radio amateurs were using such receivers for global communications with transmitter powers of perhaps only 100 W and receivers using a single device having a g_m of less than 1 mS. An interesting paper by D.G. Tucker on the history of this type of circuit is in *Radio and Electr. Eng.*, **42**, 69–80, 1972.

9 As noted above in note 8, a highly skilled operator would have found a regenerative radio receiver perfectly usable in the earlier part of this century. In that case it was really the circuit–operator combination which was usable.

10 Some very effective methods of using computers to display the relative sensitivity of the components in a circuit have been developed by R. Spence and his co-workers: *Computer Aided Design*, **8**, 49–53, January 1976.

11 Here we are following R.A. Santilli, *IEEE Trans.*, **BTR-13**, 113–18, 1967. The book *Transistors and Active Circuits* by J.G. Linvill and J.F. Gibbons, McGraw-Hill, New York, 1961, gives a more detailed discussion and has been used as a model by the majority of later authors.

This is also true of the classic work by A.P. Stern, *Proc. IRE*, **45**, 335–43, March 1957. All these authors consider a general case in which all the y parameters are complex. We have been able to use a particularly simple representation here, y_{12} imaginary and y_{21} real, because we have chosen a fairly low frequency, 10 MHz, for our experiments.

12 RCA CA3028A/B: Data Bulletin File No. 382.

13 RCA 3N187: Data Bulletin File No. 326.

14 RCA AN4431, *RF Applications of the Dual Gate MOS FET up to 500 MHz*, L.S. Baar.

15 RCA ICAN5337, *Application of the RCA CA3028A and CA3028B Integrated Circuit RF Amplifiers in the HF and VHF Ranges*, H.M. Kleinman.

3

Operational Amplifiers

3.1 Definitions

Operational amplifiers are amplifiers which have a balanced input, that is two input terminals which both have a high impedance to ground and also a high impedance between them, and a single-ended output. The gain of the amplifier is usually very high at zero frequency and is constant up to some fairly modest frequency, in the case of general purpose operational amplifiers, or to quite a high frequency in the case of a wide-band operational amplifier.

This is summarised in Fig. 3.1 where the usual symbol for the operational amplifier is shown as Fig. 3.1(*a*) and a typical frequency response, for a wide-band operational amplifier [1], is sketched in Fig. 3.1(*b*). These properties allow the operational amplifier to be used with a variety of feedback networks to provide a number of useful analog signal processing functions.

Fig. 3.1.

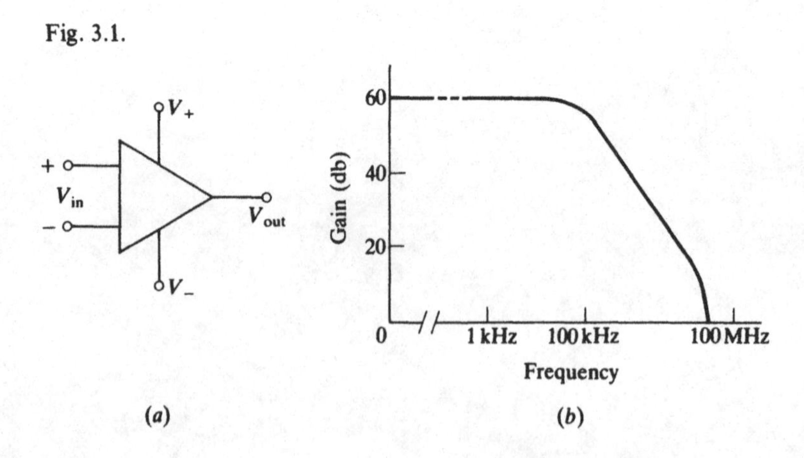

There are several excellent texts on the use, and on the design and synthesis, of operational amplifiers [2]. In this chapter we are only concerned with the design of the internal circuit. Chapter 4 is concerned with the systems which can be built up from combining several operational amplifiers. Internally, because the operational amplifier is nearly always a monolithic silicon integrated circuit, we find some of the most interesting advances in electronic circuit design, that is in new circuit shapes, which have come about during the past twenty years. However, let us look at the problem from a more classical point of view first.

3.2 A simple experimental operational amplifier circuit

To illustrate the difficulties of operational amplifier design, let us build a very simple discrete component amplifier. The proposed circuit shape is shown in Fig. 3.2.

The input stage for this amplifier, Q_1–Q_2, is based upon Blumlein's 'long tailed pair' which was discussed in connection with Fig. 2.2 in the last chapter. This gives a balanced input, but the balance is not quite perfect in Fig. 3.2 because we have decided to go from balanced to single ended circuits immediately after the input stage and this means that the currents in Q_1 and Q_2 can never be exactly matched. There are a number of other difficulties with this circuit: it will amplify common mode signals, its gain depends upon the d.c. level at the input terminals, and it has very poor dynamic range, to mention just a few. But these poor features make the circuit a good one to build and experiment with so that it becomes clear why we do not normally use circuits that look like this. Full constructional

Fig. 3.2.

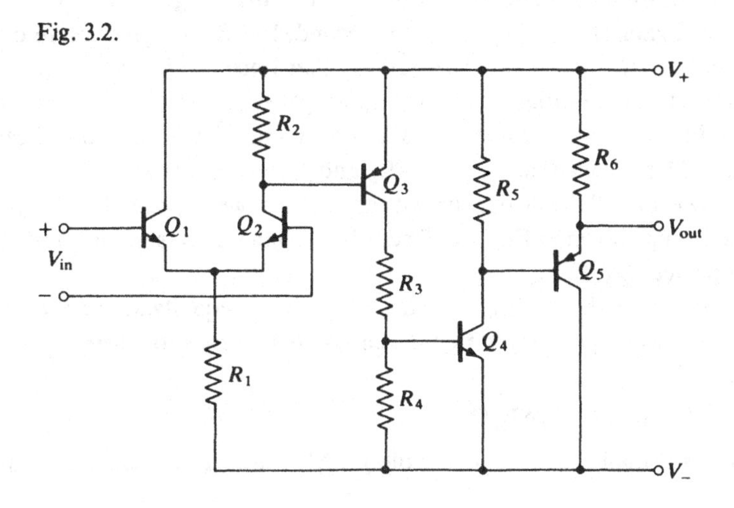

details of Fig. 3.2 are given in the Appendix and the experimental work introduces the problems of input bias current, offsets, common mode gain, output dynamic range, frequency compensation for stable operation as a feedback amplifier, slew rate, and non-linear behaviour with large output signal level at high frequency.

3.3 Circuit details of the experimental amplifier

Let us now look at how the device currents are defined in Fig. 3.2 and also calculate the gain of the circuit.

If the two input terminals in Fig. 3.2 are both held at zero potential, the sum of the collector currents in Q_1 and Q_2, $I_{C1}+I_{C2}$, is simply $(|V_-|-|V_{BE}|)/R_1$ [3]. I_{C2}, however, is also well defined because the current flowing in R_2 must be V_{BE3}/R_2. This idea of defining a current in a resistor by connecting it in parallel with the emitter junction of a transistor is a very useful one: the current in Q_3 has been defined in this way too and is $I_{C3} = V_{BE4}/R_4$. For silicon, V_{BE} is always close to 700 mV.

Note that the problem of *level shifting* is solved in this design by alternating the polarity of the bipolar transistors, npn to pnp, as we go from stage to stage. This elegant solution is quite a natural choice in a discrete component circuit [4] but in the monolithic circuits which will be discussed in the following sections, level shifting may be a problem because, in the cheapest integrated circuit processes, high quality devices of only one polarity will be available. More expensive processes allow a mixture: for example, high quality npn bipolar and p-channel MOS transistors.

Now the currents in Q_4 and Q_5 of Fig. 3.2 are not defined until we connect the amplifier up as a negative feedback amplifier. Two familiar configurations are shown in Fig. 3.3. The first, Fig. 3.3(*a*), gives us an overall gain of $-R_2/R_1$, while the second, Fig. 3.3(*b*), gives us an overall gain of $+(R_2+R_1)/R_1$. The point is that once the feedback network is connected, the output terminal of the amplifier, under the zero input signal conditions we are considering, takes up a voltage level very close to zero as well. This means that $I_{C5} = V_+/R_6$ and $I_{C4} = (V_+ + V_{BE5})/R_5$.

Now that all the device currents are defined, we can calculate the gain of the circuit shown in Fig. 3.2. To do this we ignore R_3, since it is only there to protect Q_3.

First, note that the input voltage, V_{in}, divides equally across the emitter junctions of Q_1 and Q_2 so that the change in the collector current of Q_2 will be

$$\Delta I_{C2} = g_{m2}\Delta V_{\text{in}}/2 \tag{3.1}$$

due to a small change in input voltage, ΔV_{in}. Most of this change in current will be in the base lead of Q_3, simply because we shall choose to make I_{C3} about ten times I_{C2}, which means that Q_3 is current fed and its change in collector current will be $h_{fe3}\Delta I_{C2}$. The same argument applies to Q_4; again we would make I_{C4} about ten times I_{C3}. It follows that the gain up to the collector of Q_4, and thus the overall gain because Q_5 is simply an emitter follower output stage, will be of the order of

$$G = h_{fe4}h_{fe3}g_{m2}R_5/2 \tag{3.2}$$

and as g_{m2} is $I_{C2}/(kT/e)$ or $V_{BE3}/(kT/e)R_2$, we have

$$G = h_{fe4}h_{fe3}V_{BE3}R_5/2(kT/e)R_2 \tag{3.3}$$

Now equation (3.3) gives a gain of about 25 000 for the component values detailed in the Appendix, assuming a rather modest current gain of 100 for Q_4 and Q_3. The circuit is a very instructive one to experiment with: it needs proper frequency compensation in order to be stable as a feedback amplifier, and illustrates all kinds of defects which are overcome in the more advanced circuits which have evolved and which are discussed in the following sections.

3.4 Transfer from a differential input to a single-ended output

Fig. 3.1(*a*) emphasises the feature, common to all operational amplifiers, of a balanced, or differential, input and a single-ended output. Somewhere in the circuit of the amplifier the transition must be made from balanced to single-ended circuit shape, and, for example, in our very simple amplifier of Fig. 3.2, this transition is made as soon as possible by using only one side of the input pair, Q_1 and Q_2, as an output to the following stages of amplification.

Fig. 3.3.

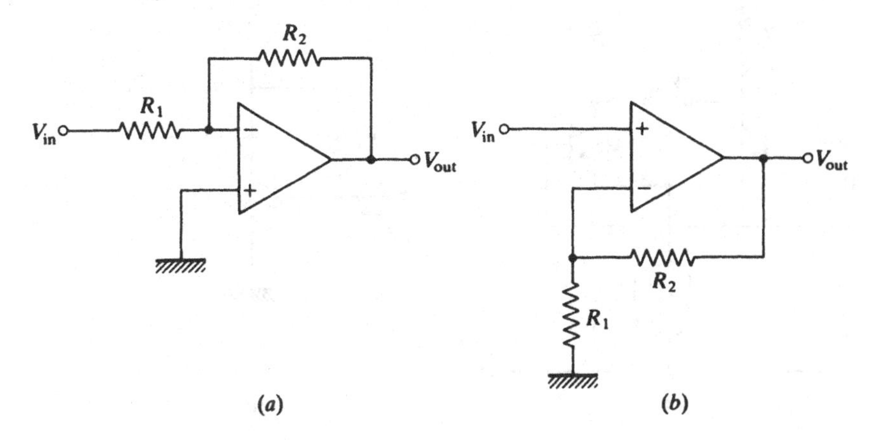

One very interesting and elegant solution to this problem was described by R.J. Widlar in a patent which was filed early in 1965 [5]. This early work is of interest because it represents one of the first steps in a real understanding of the implications of moving from electronic circuits made out of discrete transistors and other components, into the new medium of monolithic silicon. Although he himself did not express it this way, it seems that Widlar discovered a new way of inventing new circuit shapes: by embedding one familiar circuit into another. This is a big step forward from simply adding circuits together: the technique put forward in Section 2.3. Fig. 3.4 begins to explain what is meant.

In Fig. 3.4(*a*) we see the Blumlein long tailed pair again, now drawn as a symmetrical balanced input, balanced output, amplifier and with an ideal current sink, I_S, as its 'long tail'. In Fig. 3.4(*b*) we see another classical circuit [6], usually referred to as the shunt feedback circuit, which has the property of being a current to voltage converter: a change in the input current, ΔI_{in}, causes a change in the output voltage, $-\Delta V_{out} = \Delta I_{in} R_3$. The signs mean that an increase of current into the circuit produces a drop in the voltage at the collector of Q_3.

Widlar embedded Fig. 3.4(*b*) into Fig. 3.4(*a*) to produce a new circuit shape, which is shown in Fig. 3.5. In the new circuit, R_1 and R_4 become one and the same resistor, which we label R_1, we disconnect the long tailed pair from the V_+ line, its positive supply now comes through R_3, and we add an output stage, Q_4–R_5, to make the circuit symmetrical and have a single-ended output.

Fig. 3.4.

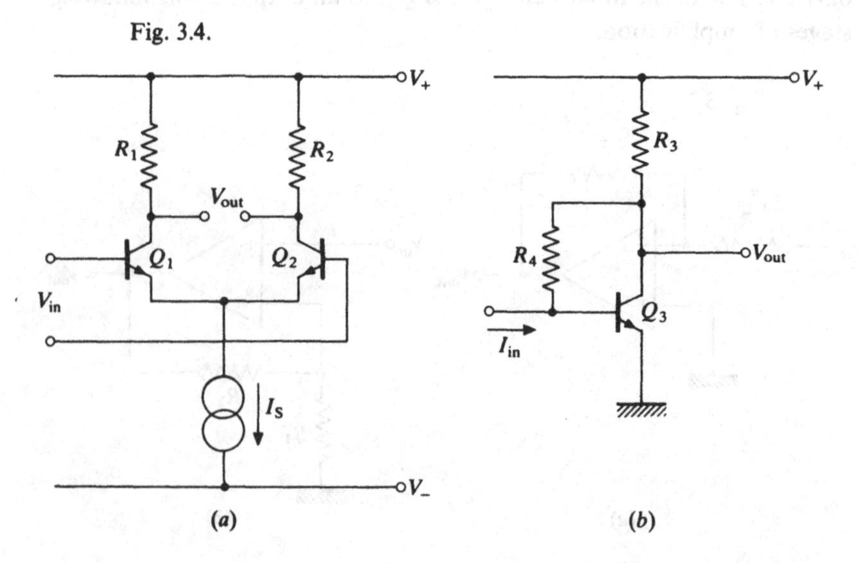

Now Fig. 3.5 is going to be a silicon integrated circuit, and this means that we may assume all four transistors to be virtually identical and, also, that we can make resistors R_1 and R_2 virtually identical. Although we shall not be able to make the absolute values of any resistors very accurate, the monolithic technique means that resistors intended to be identical will, indeed, be very closely matched. On top of all this, any change in circuit temperature, brought about either by circuit dissipation or by ambient changes, will be common to all transistors and resistors in the circuit, simply because of its small size and the fact that it is truly monolithic: all made from one small piece of silicon.

It follows that we can make R_1 and R_2 have identical values and also make Q_3 and Q_4 have identical properties. This will mean that Q_3 and Q_4 must draw identical collector currents, always assuming that R_3 and R_5 are not so large that these transistors saturate, simply because the emitters of Q_3 and Q_4 are connected to the same point (ground) and the bases of Q_3 and Q_4 are both connected to the same point through identical resistors. This property of the circuit was the central issue of Widlar's patent [5].

What is this current that flows in both Q_3 and Q_4? The answer is given when we realise that Q_3 has clamped the collector of Q_1 at $+V_{BE3}$ above ground. Consequently, neglecting the base currents of Q_3 and Q_4, we have the bottom of R_3 at $V_{BE3} + I_S R_1/2$. Because the top of R_3 is at V_+, the

Fig. 3.5.

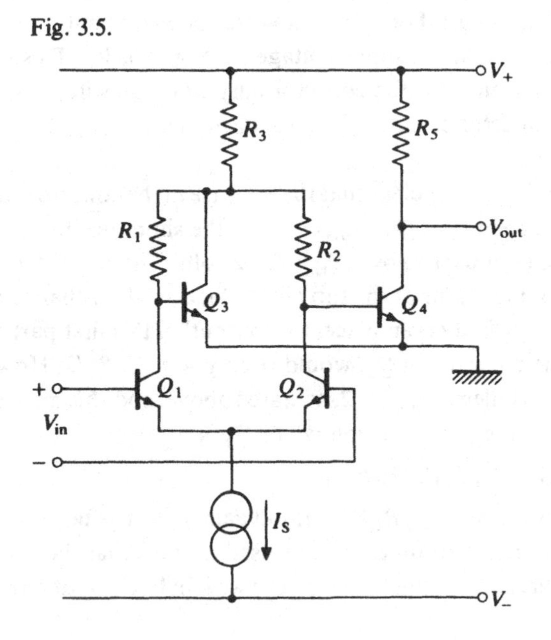

current in R_3 is now defined as

$$I_{R3} = (V_+ - V_{BE3} - I_S R_1/2)/R_3 \qquad (3.4)$$

and the collector current of Q_3, and thus of Q_4, must be $I_{R3} - I_S$, where I_S is the current sink used to bias Q_1 and Q_2. If a simple resistor, R_6, had been used for this, I_S would have been approximately equal to V_-/R_6 in magnitude.

So, provided the input terminals of the circuit shown in Fig. 3.5 are always kept close to ground potential, all the currents in the circuit are well defined and all the resistor values can be calculated. What we have not yet looked at, however, is how Widlar's new circuit shape has achieved the transition from the balanced pair of input terminals shown in Fig. 3.5, to the single output terminal at the collector of Q_4. What advantage has this new circuit over our crude discrete circuit of Fig. 3.2?

The first point about Fig. 3.5 is that it is balanced. Q_1 and Q_2 can have identical collector currents and collector–emitter voltages. This means, because Q_1 and Q_2 can be made virtually identical in the integrated circuit process, that we should be able to make the input offset voltages and currents very small indeed over a wide range of temperature.

The second point about Fig. 3.5 is that all the *differential* gain of Q_1 and Q_2 is added to the total gain of the circuit, whereas only half of this potential gain could be used in Fig. 3.2, while the *common mode* gain of Q_1 and Q_2 is effectively reduced to zero. All this is done by means of the current to voltage conversion properties of Q_3: more precisely, the properties of the circuit shown in Fig. 3.4(*b*). Let us see how this comes about.

Consider a small increase in input voltage for Fig. 3.5, V_{in}. This will be shared equally by Q_1 and Q_2: the collector current of Q_1 will *increase* by $g_m V_{in}/2$ while the collector current of Q_2 will *decrease* by $g_m V_{in}/2$, g_m being equal to $I_S/2(kT/e)$.

Referring to Fig. 3.4(*b*), it is clear that the increase in the collector current of Q_1 is an input current change, $-g_m V_{in}/2$, for the shunt feedback circuit and the collector of Q_3 must rise by $+g_m V_{in} R_1/2$ volts. The same increase in voltage must, of course, occur at the top of R_2. So, even if no change in the collector current of Q_2 had taken place, the 'output' of this first part of the circuit, which is the collector of Q_2, would rise by $+g_m V_{in} R_1/2$. However, the current in R_2 has fallen by $g_m V_{in}/2$, as stated above, and this means that the total increase in the collector voltage of Q_2 is

$$\Delta V_{C2} = \Delta V_{B4} = g_m(V_{in}/2)(R_1 + R_2) \qquad (3.5)$$

and as $R_1 = R_2$ this is simply $g_m R_1 V_{in}$: the full gain $g_m R_1$ has been realised.

Now precisely the same argument can be used to show that the common mode gain of this circuit is negligible. Suppose we join both input terminals together and apply a fairly high frequency input signal, a frequency high enough to make the current sink I_S have a finite impedance so that both Q_1 and Q_2 see a small increase in base–emitter voltage, V_{be}. Again, the top of R_2 will be lifted up a voltage $g_m V_{be} R_1$ by the current to voltage conversion properties of Q_3. The collector current of Q_2, however, has also increased so that, in place of equation (3.5) we have

$$\Delta V_{C2} = \Delta V_{B4} = g_m V_{be}(R_1 - R_2) \tag{3.6}$$

and as $R_1 = R_2$ this can be made effectively zero.

It follows that the circuit shown in Fig. 3.5 has the potentially very high differential voltage gain

$$A_d = \frac{I_S R_1}{2(kT/e)} \times \frac{I_{C4} R_5}{(kT/e)} \tag{3.7}$$

which could easily be well over 10^4, because $I_S R_1$ and $I_{C4} R_5$ can both be made a few volts whereas kT/e is only 25 mV. The common mode gain of the circuit, on the other hand, can be made very low even at quite high frequencies where it would be impossible to realise the ideal current sink, I_S, because of the shunt capacitance that must be associated with it.

3.5 The problem of large common mode signals

Fig. 3.5 is certainly a very interesting circuit shape and has been discussed at some length here because it seems to illustrate the process of circuit invention, which is really the main topic of this book, particularly well. The circuit does, however, suffer from a serious drawback in that the two input terminals must always be kept very close to, or below, ground potential simply because the collector of Q_1 has been clamped at only about 700 mV above ground [7]. This was overcome in the version of the circuit which was finally adopted by a very large number of manufacturers as their first operational amplifier in monolithic silicon [8]. The solution was to return the emitters of Q_3 and Q_4, in Fig. 3.5, to a point which is connected almost directly to the output terminal of the amplifier. Because of the way the amplifier is used, this will ensure that quite a wide range of common mode input voltage can be accommodated. It is well worthwhile studying the old 709 circuit to see how it evolved from Fig. 3.5. Q_3 and Q_4 have both been replaced with so-called Darlington pairs [9] and both have their outputs buffered with emitter followers. This design allowed both input terminals to be raised or lowered to within a few volts of the ± 15 V supply rails.

Later operational amplifier designs made an even greater range of common mode input voltage possible and this came about by the development of the part of the circuit, shown in Fig. 3.5, which has been represented up to now by the current sink I_S. What do we really use for this?

3.6 Current sources and sinks in monolithic silicon

The input transistors of an operational amplifier, for example Q_1 and Q_2 in Fig. 3.5, are quite often biassed to have collector currents as small as 10 μA in order to increase the input impedance and reduce the input bias currents of the amplifier as much as possible. We shall see in a later section of this chapter that noise considerations may suggest that higher collector currents should be used in the input transistors.

If very low collector currents are used, however, a very beautiful circuit idea may be used to determine them. This idea, again due to R.J. Widlar [10], is shown in Fig. 3.6 and has become well known as the Widlar current source.

The principle of the circuit is that Q_1 is clamped active by connecting its base to its collector. Its collector current will be $(V_+ + |V_-| - V_{BE1})/R_1$ and is thus very well defined. Simple transistor theory tells us that we than have

$$(V_+ + |V_-| - V_{BE1})/R_1 = I_{CBO} \exp[V_{BE1}/(kT/e)] \tag{3.8}$$

where we neglect the base currents of both Q_1 and Q_2 compared to the collector current of Q_1.

Q_2 will pass a smaller current than that given by equation (3.8) because its base–emitter voltage is smaller than V_{BE1} by an amount $I_S R_2$. We are, of course, assuming Q_1 and Q_2 to be identical and to have the same value of I_{CBO}. This means that

$$I_S = I_{CBO} \exp[(V_{BE1} - I_S R_2)/(kT/e)] \tag{3.9}$$

or, dividing (3.9) by (3.8),

$$I_S = [(V_+ + |V_-| - V_{BE1})/R_1] \exp[-I_S R_2/(kT/e)] \tag{3.10}$$

Fig. 3.6.

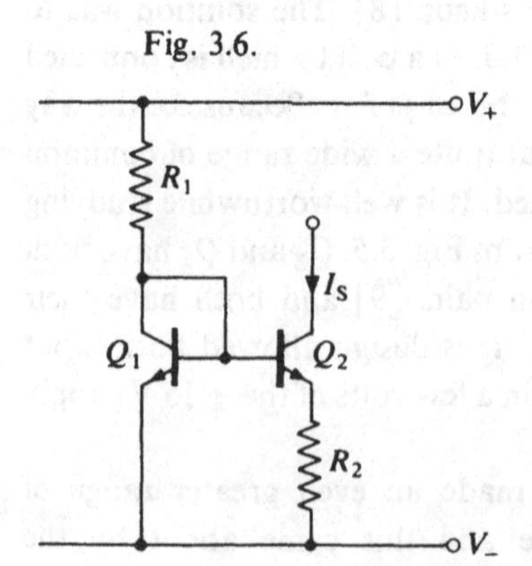

Equation (3.10) can only be solved for I_S numerically; it is a transcendental equation, but it is quite simple to see what is involved. If we needed a current sink of 15 μA, a sensible design would be to use 10 kΩ for R_1, as this is about the largest resistor value we would care to have in a simple integrated circuit, and take the top of R_1 to ground. The collector current for Q_1 would then be just below 1.5 mA for a -15 V negative supply. By making $I_S R_2/(kT/e)$ equal to 4.6, because $\exp(-4.6)$ is 10^{-2}, we would have $I_S = 15\,\mu$A. As (kT/e) is about 25 mV, this means $I_S R_2$ would be 115 mV and R_2 must be 7.67 kΩ.

This circuit idea will only work in an integrated circuit, or, of course, with a true dual transistor, because it relies upon Q_1 and Q_2 not only being identical but also being at the same temperature. Only then may we assume that the I_{CBO}, the collector junction reverse leakage current, is the same in both devices.

3.7 The current mirror

If we make $R_2 = 0$ in Fig. 3.6, equation (3.10) tells us that I_S is now identical to the current determined by R_1 and the supply voltages. Looked at in detail, we have the well known current mirror circuit shown in Fig. 3.7, in which I_2 always equals I_1.

It is interesting to note that this circuit, although it looks much simpler than Fig. 3.6, began to be used in monolithic operational amplifier designs only after some years had passed. The current mirror of Fig. 3.7 is a really new idea in electronics: a really new circuit shape which takes a little getting used to. Fig. 3.7 is the simplest of all its possible forms [11]. There are modifications which increase the impedance of the output, and improve the accuracy of the $I_1 = I_2$ condition, and further circuit modifications which allow excellent current sources to be made with the very low current gain pnp devices produced by the cheapest bipolar integrated circuit process techniques. The current mirror idea can also be realised using MOS devices [12].

Fig. 3.7.

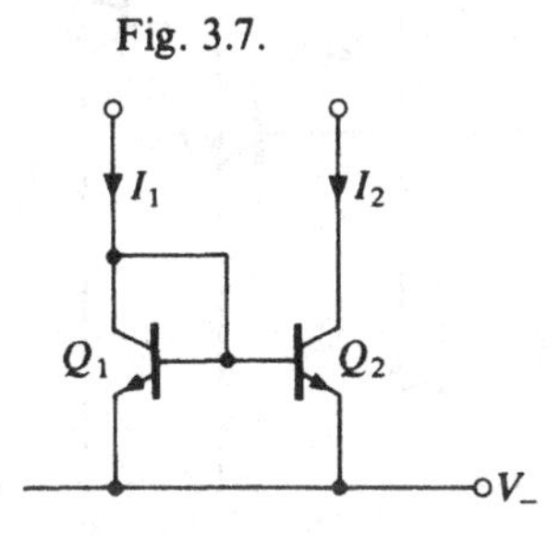

The current mirror is an excellent solution to the problem discussed in Section 3.4; going from a differential input to a single-ended output. Fig. 3.8 shows a circuit shape for an operational amplifier which illustratees this.

In this design, the decision to move into single-ended circuits has been postponed until the third stage of amplification. Two balanced stages are used to begin with; the first stage is Q_1 and Q_2 with resistive loads and the Widlar current sink of Fig. 3.6 as the 'long tail', because the input stage would operate at very low collector current in order to obtain high input impedance and low input bias current. The second stage, Q_3 and Q_4, has a simple current mirror, Q_{11} and Q_{12}, as its 'long tail': this time a current source because we have changed device polarity in going from stage one to stage two. It is at this point in the circuit where we would probably use one of the modified current sources mentioned above [11] and find MOS transistors being used for Q_3 and Q_4 [12].

The most important feature of the second stage is the pair, Q_5 and Q_6. These two devices form a current mirror load which allows a very effective transition to the single-ended output stages, Q_7 and Q_8. The point is that Q_6 mirrors the collector current of Q_3 so that the full differential *current* output of Q_3 and Q_4, that is $I_{C4} - I_{C3}$, is passed into the base of Q_7. This circuit shape, the shape formed by Q_3, Q_4, Q_5, Q_6, and Q_7, is a really new step forward in electronic circuit design: an amplifier circuit entirely composed of transistors.

As a circuit shape, the first three stages of the amplifier shown in Fig. 3.8 have been adopted in a number of operational amplifier designs: for

Fig. 3.8.

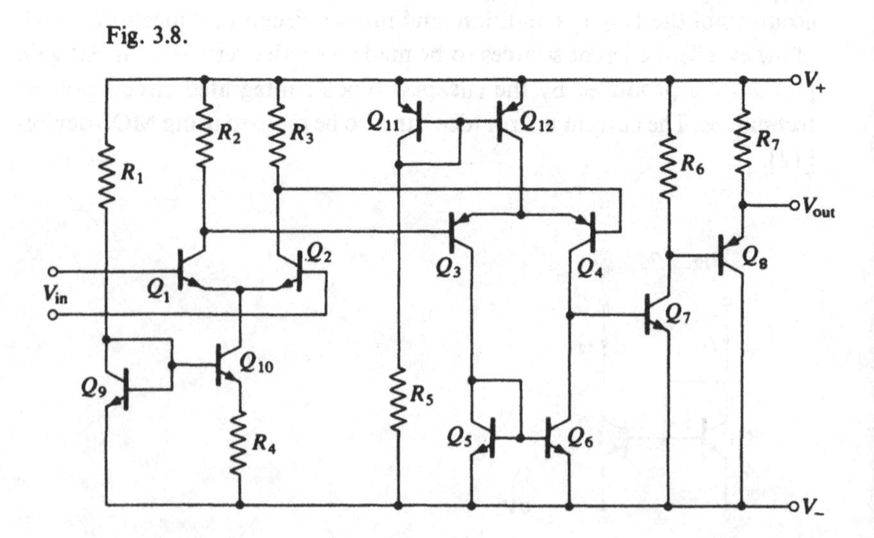

example, the OP–05 and OP–07 precision bipolar devices [13] and the CA3100 BiMOS wide-band device [14]. The output stage shown in Fig. 3.8 is a drastic oversimplification because it is simply the same one used in Fig. 3.2. This is because output stages have their own special problems and are considered in Chapter 9.

3.8 Input bias current cancellation

To conclude this chapter on operational amplifiers we look at just one more development which has come about at the input end: the idea of input bias current cancellation. This leads to some very interesting circuit shapes which are particular for monolithic circuits.

Operational amplifiers which use bipolar transistors as their input devices can have extremely small input offset voltages and currents: as small as 10 μV and 0.5 nA. This is in contrast to amplifiers which use field effect transistors where the input offset voltages may be very large: as large as 10 mV. The disadvantage of the bipolar input device, however, is its input bias current. This can be made as low as 50 nA [15] by using devices with very high current gain at very low collector current, but noise considerations and processing problems [16] suggest that some kind of input bias cancellation technique might be a better solution.

A very elegant method of input bias current cancellation was described, almost simultaneously, by Kuijk [17] and by del Corso and Pozzolo [18]. The essential idea is shown in Fig. 3.9.

In Fig. 3.9, we see Q_1 as one of the input transistors of the amplifier in just the same way as Q_1 was one of the input transistors in our previous circuit, Fig. 3.8. In series with Q_1 we put an identical transistor, Q_{11}, which must

Fig. 3.9.

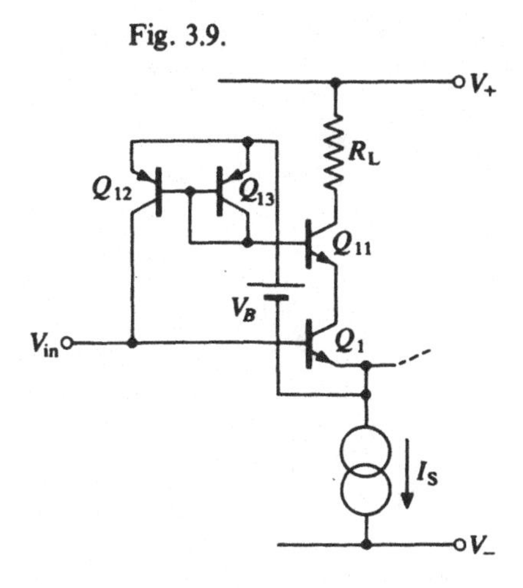

take the same base current as Q_1, provided we ensure that it is always active, and this base current is supplied by Q_{13} from a floating supply, V_B.

Now Q_{13} forms a current mirror with Q_{12}. It follows that the same base current that is flowing, at any time, into Q_{11} will be supplied by Q_{12} to Q_1 and, if the current mirror Q_{12} and Q_{13} could be perfect, and if Q_{11} and Q_1 could be perfectly matched, zero bias current would be needed in the true input lead going to the terminal marked 'V_{in}' in Fig. 3.9.

This input bias current compensation idea is not simply a constant compensation idea. It is, in fact, positive feedback, and the small signal bias current should also be cancelled out giving very high input impedance and problems of instability. This is dealt with in detail by del Corso and Pozzolo [18]. Here, we are concerned only with the circuit shape idea and the use of the current mirror in this novel way. The floating supply, V_B, which is needed in Fig. 3.9 is also realised by means of current mirrors and this is shown in Fig. 3.10. Two current mirrors are used to bias four diodes D_1–D_4 with a forward current approximately equal to $(V_+ + |V_-|)/R$. Because this is a silicon circuit, V_B will be just below 2.8 V and this is quite enough to ensure that Q_1 and Q_{11}, in Fig. 3.9, are active over quite a wide range of common mode input voltage.

The circuit shape ideas shown in Figs. 3.9 and 3.10 were applied with great success in the well known OP–07 and OP–05 operational amplifiers [13, 16].

A completely different circuit shape idea for input bias current compensation, but still of the positive feedback kind aiming at complete cancellation, is shown in Fig. 3.11. This idea is found in the CA3193 and

Fig. 3.10.

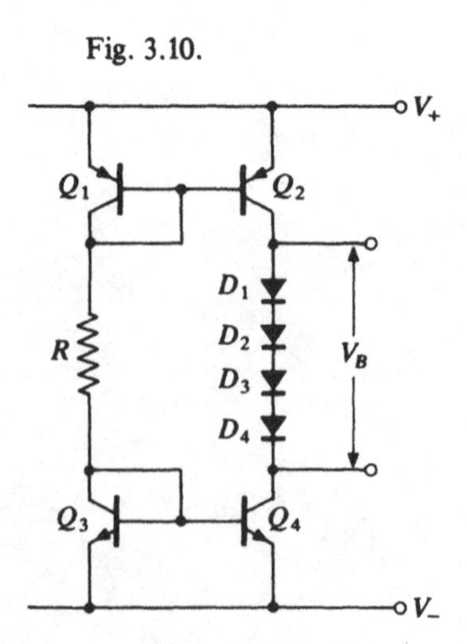

CA3493 precision operational amplifiers [19]. No current mirrors are used in this case, except to realise the floating supply, V_B, which is needed, again along the lines shown in Fig. 3.10. Fig. 3.11 relies on symmetry and the use of pnp and npn bipolar devices, and is a beautiful example of the circuit designer's art. Once seen, Fig. 3.11 is such an obvious solution, but to get started on it the first step was, as in Fig. 3.9, the apparently trivial step of putting Q_{11} in series with Q_1 and so obtaining a second base current equal in magnitude to the base current we wish to cancel out. In Fig. 3.9, this second base current is reversed in direction and reproduced at the voltage level required by means of the current mirror, Q_{12} and Q_{13}. In Fig. 3.11, this reversal and level shifting is done by a Q_{12} and Q_{13} which do not mirror the current but mirror Q_1 and Q_{11} in symmetry and polarity. This circuit is dealt with in detail by Laude [20].

Finally, Fig. 3.12 shows the input bias current cancellation scheme used in the low noise amplifiers OP–27 and OP–37 [21]. In this design, only the d.c. level of the input bias current is cancelled out. There is no positive feedback, so there is no problem with stability, and the noise performance is better because the cancellation current is no longer correlated with the input bias current, as it was in both the previous circuits because it was derived from the same source. In Fig. 3.12 the cancellation current is derived from the base of Q_{11}, which is not in series with Q_1 anymore but arranged to draw the same direct current as Q_1 by virtue of the multiple current sink Q_{15}, Q_{16}, Q_{17}, and Q_{18}: all these devices are identical. The base current of Q_{11} is again mirrored by Q_{12} and Q_{13} and supplied to the base of Q_1, just as it was in Fig. 3.9. Q_{14} is a very neat feature of this circuit: it is

Fig. 3.11.

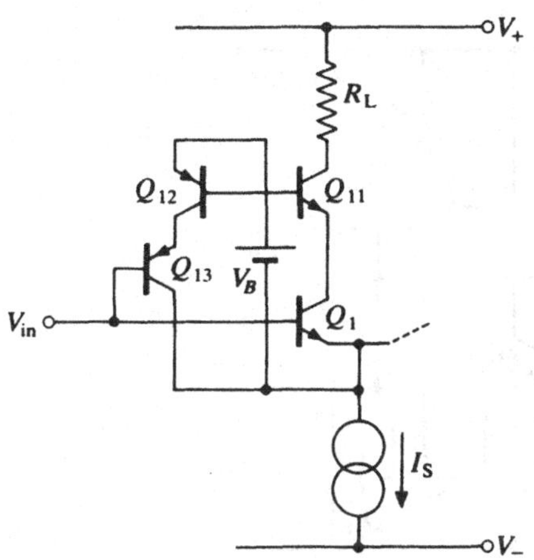

being used to make the collector–base voltage of Q_{11} always about equal to that of Q_1, no matter what d.c. level is put on the input terminal. This circuit is dealt with in detail by Erdi [16].

3.9 Reflections

In this chapter we have looked at the changes which have taken place in operational amplifier circuit design, in the circuit shapes being used, over a period of nearly twenty years. The extent of this change is illustrated if we compare Figs. 3.2 and 3.12. All of Fig. 3.12 is involved in providing the function previously provided by just Q_1 and R_1 in Fig. 3.2. At first sight, this looks like a dramatic increase in complexity, and it is, but we must remember that it is, in fact, no more difficult to actually manufacture Fig. 3.12, in monolithic silicon, than it is to manufacture the input stage of Fig. 3.2.

What has really changed over the past twenty years, in operational amplifier design, is our way of thinking about circuits and circuit shapes. With the majority of people, it takes quite some time to develop a new way of looking at things, and this is one reason why we find one particular person dominating the field at a time of rapid change. The conservatism of the main stream is also shown up in the literature. For example, Q_1 in Figs. 3.6 and 3.7 is still repeatedly referred to as 'diode connected' in many texts, or even replaced by a diode symbol in otherwise very accurately produced data sheets. Q_1 is not operating as a diode at all: it is a transistor which is clamped active by making $V_{CB} = 0$. The current in Q_1 flows, predominantly, *into* the n-type collector and out of the n-type emitter. A diode is a pn junction where the current flows into the p side and out of the n side, and

Fig. 3.12.

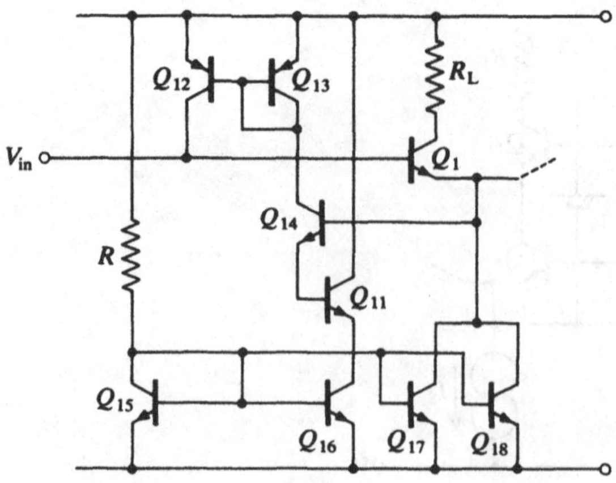

this current does not follow the right-hand side of equation (3.8) but varies as $\{\exp[V_F/(kT/e)] - 1\}$.

Finally, has this development in circuit complexity meant that experimentalists are no longer able to work with circuits like Fig. 3.12 when they have to work in simple laboratories which have no facilities for making integrated circuits? The answer to this is that it is possible to experiment with circuit shapes as complex as Fig. 3.12, which need the close matching and isothermal properties of monolithic silicon, by using dual transistors [22] and transistor arrays [23]. However, this is, perhaps, not the best way to look at the new situation, which is that new devices are available to us as experimentalists, and it is the experiments and new circuits which we can build with these that now become the main interest. This is the idea put forward in the next chapter.

Notes

1 The frequency response shown in Fig. 3.1(b) is for the RCA CA 3100, Data Bulletin File No. 625.

2 *The Art of Electronics* by P. Horowitz and W. Hill, Cambridge University Press, 1980, Ch. 3, is particularly good for operational amplifiers. Note also the *Laboratory Manual* by P. Horowitz and I. Robinson, Cambridge University Press, 1981, which goes with this. For detailed design and synthesis of operational amplifiers, see *Analysis and Design of Analog Integrated Circuits* by P.R. Gray and R.G. Meyer, John Wiley, 1984 (2nd edition). More references are given here in Chapter 4, note 1.

3 In all the algebra here we use I for current and V for voltage with upper case suffices (E, B, C for emitter, base, collector) to denote d.c. levels and lower case for small signal. The numbers following the suffix indicate the device in the circuit.

4 It is interesting to note that such an elegant solution was not possible in the older electronics. With high vacuum, hot emitter, devices only one polarity of charge carrier is possible in practice.

5 US Patent No. 3364434 of 16 January, 1968, to R.J. Widlar and the Fairchild Co.

6 M.G. Scroggie, *Wireless World*, **51**, 194–6, July 1945.

7 Despite this, the circuit was used in the Fairchild 702A, which was the very first silicon integrated operational amplifier circuit to become easily available for experimenters to work with. The circuit was, apparently, not described in the usual journals but in a Fairchild application report: AR131 dated January 1965, written by R.J. Widlar.

8 This was the famous 709 circuit described by R.J. Widlar in *Proc. Nat. Electronics Conf.*, **21**, 85–9, 1965.

9 US Patent No. 2663806 of 22 December 1953, to S. Darlington and Bell Labs. Inc.

10 US Patent No. 3320439 of 16 May 1967, to R.J. Widlar and the Fairchild Co.

11 *Analysis and Design of Analog Integrated Circuits*, by P.R. Gray and R.G. Meyer, John Wiley, 1984, has chapter 4 on the design of the various types of current sources, sinks and mirrors.
12 See the two papers by O.H. Schade, *RCA Review*, **37**, 404–24, 1976, and **39**, 250–77, 1978.
13 Precision Monolithics Inc. Data book, 1984, pp. 5-33 and 5-50.
14 See note 1.
15 R.J. Widlar, National Semiconductor Corp. Application Note AN–29, 1969.
16 G. Erdi, *IEEE J. Solid State Circuits*, **SC–16**, 653–61, 1981.
17 K.E. Kuijk, *IEEE J. Solid State Circuits*, **SC–8**, 458–62, 1973.
18 D. del Corso and V. Pozzolo, *Alta Frequenza*, **XLIII**, No. 12, 1018–22, 1974.
19 RCA CA3193, Data Bulletin File No. 1249. RCA CA3493, Data Bulletin File No. 1290.
20 D.P. Laude, *IEEE J. Solid State Circuits*, **SC–16**, 748–50, 1981.
21 Precision Monolithics Inc. Data Book, 1984, pp. 5-127 and 5-139.
22 Precision Monolithics Inc. Data Book, 1984, Section 9.
23 Chapter 1, note 11.

4

Operational Amplifier Systems

4.1 Systems

Good quality operational amplifiers, produced as cheap monolithic silicon integrated circuits, became available in about 1967. From that time onwards it was possible to construct quite complicated analog signal processing systems, involving well over 100 active devices, in a matter of a few hours. There are several excellent texts which deal with this [1], usually in combination with a detailed study of the internal design of the operational amplifier itself. Here, we attempt to see some continuity in the circuit shape and circuit design ideas, presented in previous chapters, and also look at some of the rather unexpected experimental facts which may come up when we work with operational amplifier systems.

4.2 Combining operational amplifier circuits

The idea of combining or adding circuits together to form more complicated circuits was examined in the previous three chapters. With operational amplifiers, their own internal structure being complicated enough to justify the use of the term 'system', we can certainly see examples of the same kind of procedure: adding simple systems together to form more complex and more useful systems.

Consider, as an example, Fig. 3.1, which was used here to introduce the idea of an operational amplifier in the first place. The amplifier was first presented as one having a balanced input but, moving on to Fig. 3.3, we see that this useful property of a balanced input is immediately lost when we come to use the operational amplifier. Because feedback must be used to set the gain of the amplifier, and its d.c. levels, the resulting *feedback amplifiers* have only one input terminal available. In Fig. 3.3(*a*) this is accessed through R_1, and the input impedance of the feedback amplifier is thus equal to R_1, and probably quite low. In Fig. 3.3(*b*) the positive input terminal of the original operational amplifier is available, and the input impedance of the feedback amplifier may be very high: the original Z_{in} of the operational amplifier multiplied by the loop gain of the feedback amplifier.

Fig. 4.1 shows a very simple way in which a feedback amplifier might be made, using an operational amplifier and just four resistors, so that we regain the possibility of a balanced input. The idea is not a very good one but it is useful to look at because it will lead to a much better idea, and also acts as a start in the theoretical approach which is needed to deal with systems built up by combining several operational amplifier circuits.

Fig. 4.1 is really an attempt to superimpose Fig. 3.3(*a*) on Fig. 3.3(*b*). If the resulting circuit is going to be a feedback amplifier with a balanced input, two conditions should be satisfied:

(*a*) The transfer function should be of the form

$$V_{out} = G(V_a - V_b) \tag{4.1}$$

(*b*) The impedance looking into terminal V_a and ground should be the same as the impedance looking into terminal V_b and ground. This is what 'balanced input' means.

To find the relative values of R_a, R_b, R_c and R_d which satisfy these two conditions, we assume that the operational amplifier has very high gain and a very high input impedance relative to the four resistors. Input bias currents and input offset voltage and current are neglected. The two operational amplifier input terminals, + and −, may then be assumed to be at the same potential, V. Nodal analysis of Fig. 4.1 then gives

$$(V_a - V)/R_a = (V - V_{out})/R_b \tag{4.2}$$

$$(V_b - V)/R_c = (V)/R_d \tag{4.3}$$

and the only way in which a relationship of the form given by equation (4.1) can be obtained from equations (4.2) and (4.3) is to make $R_a/R_b = R_c/R_d$.

Fig. 4.1.

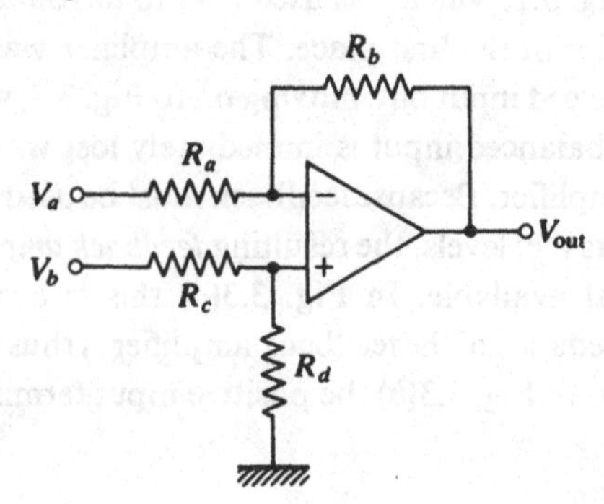

We can then subtract equation (4.3) from equation (4.2) to obtain

$$-V_{out} = (R_b/R_a)(V_a - V_b) \tag{4.4}$$

Equation (4.4) then agrees with equation (4.1). The gain, G, of the amplifier shown in Fig. 4.1 is $-R_b/R_a$. However, condition (*b*), above, will not be satisfied unless we also make $(R_c + R_d) = R_a$. The impedance looking into either input terminal will then be equal to R_a. This is not a good solution to the problem of obtaining a balanced input impedance, particularly if both high input impedance and high gain are required.

4.3 A three amplifier system

An excellent balanced input amplifier may be made up using three operational amplifiers. This design is a good illustration of adding operational amplifier circuits together and its evolution is illustrated in Fig. 4.2.

In the top left-hand corner of Fig. 4.2, we show the feedback amplifier circuit first presented here as Fig. 3.3(*b*). This has an overall gain of $V_a/V_1 = (R_1 + R_2)/R_1$. Immediately underneath this we draw the mirror image of this circuit: a feedback amplifier with an identical gain, $V_b/V_2 = (R_1 + R_2)/R_1$. On the right-hand side of Fig. 4.2, we simply repeat Fig. 4.1 with the resistors $R_a = R_c$ and $R_b = R_d$. This is the easiest way to give this circuit the transfer function $V_{out} = -(R_b/R_a)(V_a - V_b)$.

Fig. 4.2.

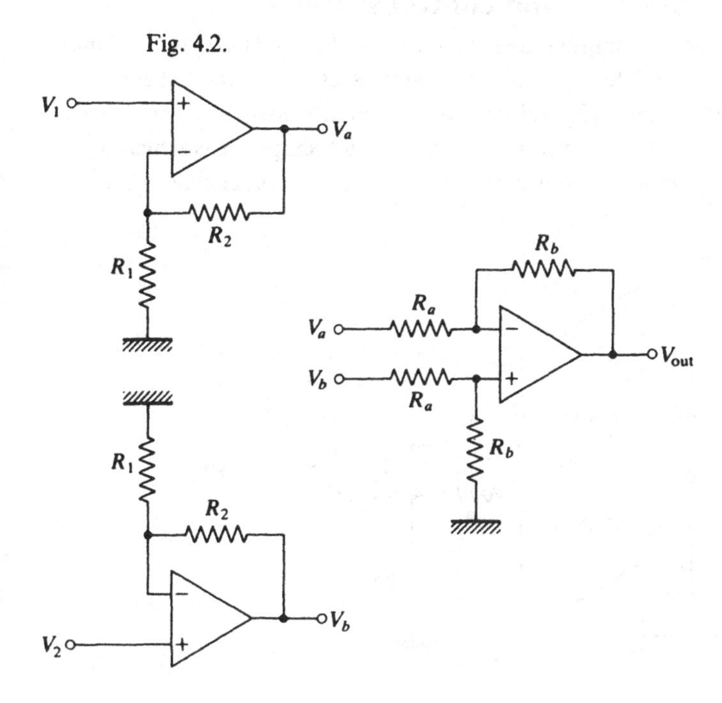

Fig. 4.3 shows the result of adding everything shown in Fig. 4.2 together. Note how the ground connection associated with the two resistors labelled 'R_1' in Fig. 4.2 disappears in Fig. 4.3. This ensures a true balance in this input stage amplifier: the currents in the two resistors labelled 'R_2' in Fig. 4.3 can only be equal in magnitude but opposite in direction. The two output voltages from the input stage, V_a and V_b, are taken directly to the input terminals of the output stage. The fact that the load seen by the two input amplifiers is different is of little importance now because the output impedance of the input amplifiers is so small.

Overall, the transfer function of the balanced amplifier shown in Fig. 4.3 is

$$V_{out}/(V_1 - V_2) = -R_b(R_1 + R_2)/R_1R_a \quad (4.5)$$

and the circuit is a very useful one for all kinds of instrumentation applications where a really balanced input, of very high input impedance, is needed. It is important to note that good performance requires very close matching between the resistors which are marked as having the same value in Fig. 4.3, and it is also essential that the two input operational amplifiers are as identical as possible. The only way to ensure this is to make use of a dual device, where two identical operational amplifiers have been made side by side on one piece of silicon [2].

4.4 Measurement and control systems

Signal processing systems, which can be built up using operational amplifiers of the kind described in the previous section, are treated at length in all the standard texts [3] and have only been discussed here very briefly in order to illustrate the continuity of the circuit shape ideas which have been presented in the previous chapters. Let us now look at some systems,

Fig. 4.3.

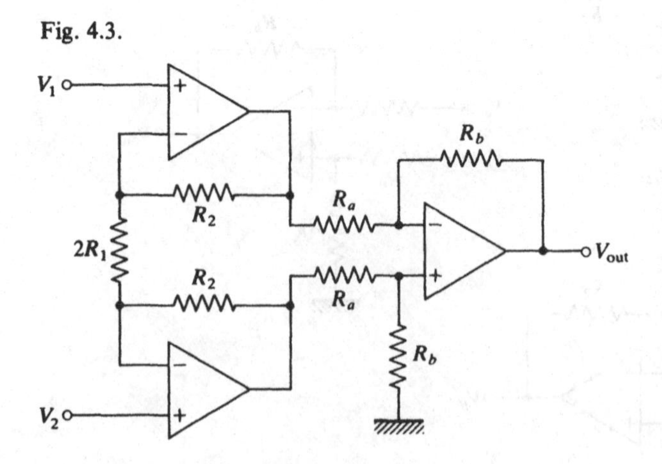

which can be built up using operational amplifiers, for measurement and control applications.

The measurement and control problem which will be taken as an example here is one which comes from the circuit design problems of the operational amplifier itself. This is the problem of measuring the open-loop gain. Consider Fig. 4.4 to begin with.

Fig. 4.4 shows what would be called an open-loop measurement system. The resistor R_3 is made as small as possible, for example 10 Ω, while resistors R_1 and R_2 are very large. The experimental procedure is to set $V_2 = 0$ and then adjust V_1 until the output voltage of the device under test (DUT) is zero. V_1 now reads the input offset voltage of the DUT, V_{os}, multiplied by the large number R_1/R_3. V_{os} can be several millivolts in the cheaper general purpose amplifiers and this means that R_1 will probably be about 10 k. It also means that V_1 will have to be a very stable voltage source indeed because the DUT may have a gain as high as 10^6. This is the first disadvantage of this open-loop system: the gain from V_1 to V_3 is quite high at about 10^3.

The second disadvantage of the open loop system becomes obvious when we try to measure the gain of the DUT by varying V_2 and observing V_3. If the gain of the DUT is as high as 10^6, R_2 would have to be a very high value: for example, 1 M or even more. If we begin our experiment by increasing V_2 from zero and observing V_3, we may well run into difficulties, quite apart from any problems with the ultra-stability required in the source V_1. V_3 may be observed to change discontinuously; in fact this

Fig. 4.4.

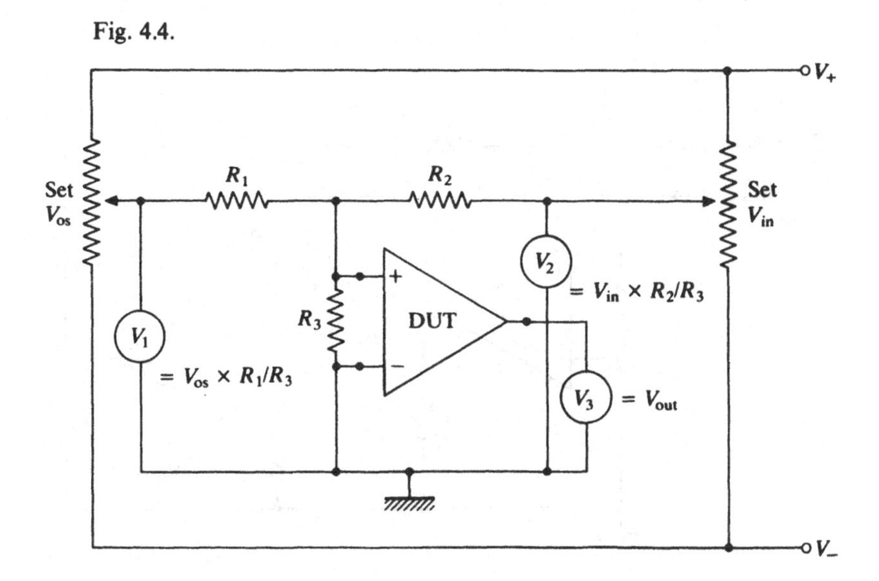

problem may well arise at the beginning of the experiment when we try to measure V_{os}. As we shall see, these difficulties arise because the output voltage of the DUT may not be a *single valued* function of the input voltage, and the way to get around this problem is to use a *closed-loop* measurement system of the kind shown in Fig. 4.5.

In Fig. 4.5, the DUT is made part of a control loop by taking the output voltage of the DUT to the input of a precision operational amplifier, A_1, connected as a unity gain feedback amplifier: $R_6/R_4 = 1$. The output of this precision amplifier then provides the input voltage for the DUT through the potential divider R_3/R_2, as in Fig. 4.4.

To measure the open-loop gain of the DUT with the set-up shown in Fig. 4.5, we introduce a second input to the amplifier A_1 via R_5. This input is labelled 'control V_{out}' in Fig. 4.5 because this is exactly what it does. If our control loop gain, that is R_3/R_2 multiplied by the DUT gain, is well over unity, say about ten, then the output voltage of the DUT will always be kept nearly equal and opposite to the 'control V_{out}' input voltage while the output voltage of the precision amplifier, A_1, will equal $V_{in}(R_2/R_3)$, where V_{in} is the true input voltage of the DUT. This assumes that the 'set V_{os}' input has been correctly adjusted to begin with so that $V_{out} = 0$ for $V_{in} = 0$.

The reasons that this closed-loop measurement and control system avoids all the problems associated with the previous open-loop system are that the gains between the various input variables and the output observables are no longer high, and also that we have now made the *output* voltage of the DUT our independent experimental variable, while the input voltage is the dependent variable. This means that the problems associated

Fig. 4.5.

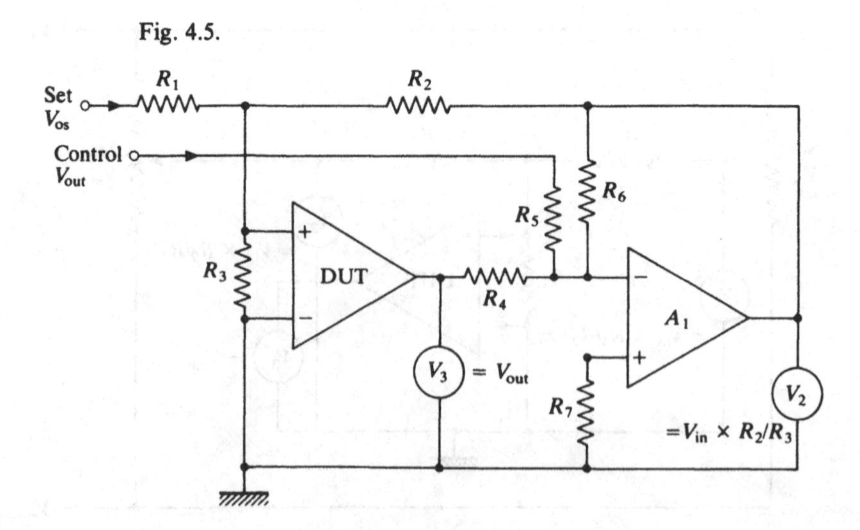

with the fact that the function $V_{out} = f(V_{in})$ may be multi-valued are avoided because the inverse function, $V_{in} = F(V_{out})$ is single valued. This is discussed in more detail in the next section.

4.5 Measurements

Fig. 4.6 shows the output voltage–input voltage characteristic of a high quality operational amplifier, measured using the closed-loop system shown in Fig. 4.5. This particular amplifier was a PMI OP–05 device working into a 30 k load. It is clear that the amplifier shows a high gain, about 350 000, with excellent linearity over an output voltage swing of ±10 V. Fig. 4.6 shows the true zero frequency open-loop gain characteristic of this OP–05 because the 'control V_{out}' input of Fig. 4.5 was adjusted slowly by hand while the characteristic shown in Fig. 4.6 was plotted on an XY recorder.

In complete contrast to the well behaved input–output characteristic shown in Fig. 4.6, Fig. 4.7 shows the characteristic of a 741 type operational amplifier, again working into a 30 k load. Although the 741 device would be considered obsolete at the time of writing, it is still a very easily available part and is still very popular. All 741 devices which follow the original design and layout show a characteristic similar to the one shown in Fig. 4.7: the amplifier has a high gain, at small output voltage levels, *but apparently of the wrong sign.* As the output voltage approaches the power supply levels, ±15 V, the gain increases in magnitude, becomes infinite, and finally takes what would be considered the correct sign over a very small range of output voltage just before saturation [4].

Fig. 4.6.

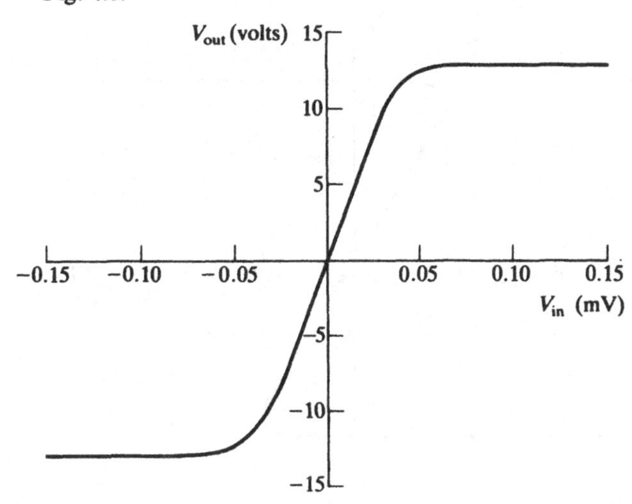

4.6 Thermal design of integrated circuits

The unexpected behaviour of the 741 operational amplifier that is shown in Fig. 4.7 is due to the poor layout of the device, as a silicon integrated circuit, from a *thermal* point of view. This is a very important aspect of integrated circuit design which has received very little attention in the literature [5]. The thermal design of an integrated circuit implies a consideration of the fact that the area covered by the circuit cannot possibly be isothermal when the circuit is working. Some regions, for example around the output transistors, must work at a higher temperature than other regions, for example around the input transistors, and this non-uniform temperature distribution will change with the output voltage level.

We can see the origin of the kind of behaviour shown in Fig. 4.7 if we go back to an earlier circuit considered in this book: Fig. 3.5. This earlier diagram is drawn again here as Fig. 4.8, but this time attention is drawn to the thermal coupling which must exist between the four transistors by drawing these all side by side as though a transistor array [6] had been used to make up the circuit. Indeed, a great deal can be learnt about thermal design and the problems of thermal coupling by building simple circuits with such arrays.

Now when the circuit shown in Fig. 4.8 was discussed in Chapter 3, as Fig. 3.5, the assumption was made that all four transistors were always at the same temperature. However, when we consider the power dissipation in these devices, it is obvious that there must be a temperature gradient along

Fig. 4.7.

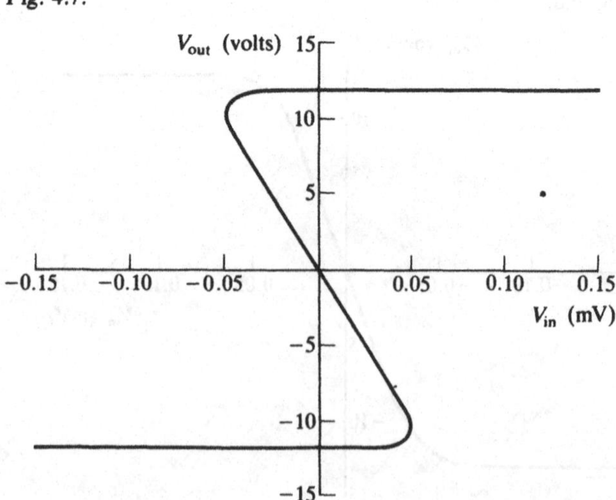

Q_1, Q_2, Q_3 and Q_4. The power dissipation in Q_1 and Q_2 will be very small while the power dissipation in Q_4 will be perhaps ten times greater. It follows that Q_2 will always be at a slightly higher temperature than Q_1 and we shall find that $V_{BE1} > V_{BE2}$, that is an effective positive input voltage will be needed to balance the currents in the two input transistors: the normal situation for zero input signal. It follows that the circuit, laid out as it is in Fig. 4.8, will show a positive offset voltage.

The power dissipation in Q_4, which is the main cause of the temperature gradient between Q_1 and Q_2, is not constant, however, but varies as

$$P_{Q4} = V_{C4}(V_+ - V_{C4})/R_5 \tag{4.6}$$

Depending upon our choice of d.c. level for V_{C4}, that is the voltage level corresponding to the input offset voltage, we may have P_{Q4} either increasing or decreasing with V_{C4}.

As an example, let us decide to make $V_{C4} = V_+/4$ correspond to our positive offset voltage V_{os} and begin to build up an input voltage–output voltage characteristic for the circuit shown in Fig. 4.8. This is done in Fig. 4.9, where we begin at the point 1 with $V_{in} = V_{os}$ and $V_{C4} = V_+/4$. Now consider moving towards a higher value of V_{C4} which, from a purely circuit connection point of view, would call for a reduction in V_{in}.

Now a reduction in V_{in}, in Fig. 4.8, means that V_{BE1} must be reduced relative to V_{BE2}. From a thermal point of view, however, an increase in V_{C4} from the $V_+/4$ level will cause an increase in the power dissipation in Q_4, as shown by equation (4.6). This may increase the temperature of Q_2, relative to Q_1, so much that the net changes in V_{BE1} and V_{BE2} add up to an increase

Fig. 4.8.

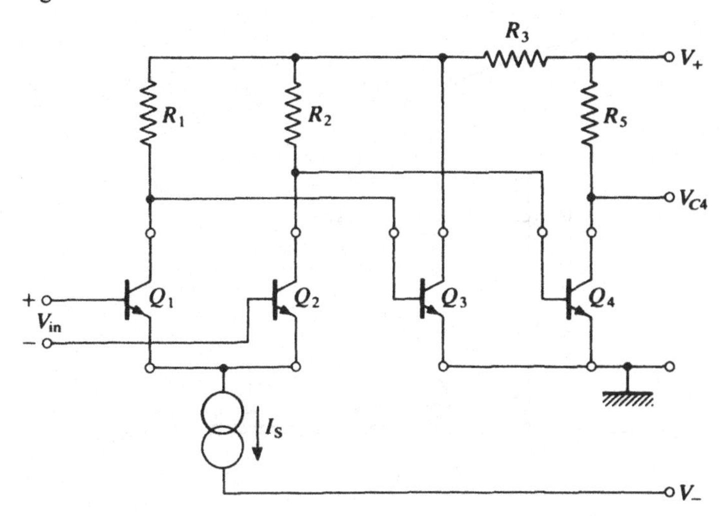

in V_{in}, as shown in Fig. 4.9. This increase will persist until the output level, V_{C4}, reaches $V_{C4} = V_+/2$, point 2 in Fig. 4.9, where differentiation of equation (4.6),

$$dP_{Q4}/dV_{C4} = (V_+ - 2V_{C4})/R_5 \tag{4.7}$$

shows that P_{Q4} has reached a maximum. Further increase in V_{C4}, towards point 3 in Fig. 4.9, will relax the thermal gradient between Q_1 and Q_2, and the input voltage will fall, as simple circuit considerations would lead us to expect.

The other half of the curve shown in Fig. 4.9, 1–4–5, may be understood by following a similar argument. As V_{C4} is reduced from the d.c. level, $V_+/4$, the power dissipation in Q_4 falls and the inequality, $V_{BE1} > V_{BE2}$, becomes less severe. This leads to an apparent drop in V_{in}: point 1 to point 4 in Fig. 4.9. However, eventually the power dissipation in Q_4 becomes negligible, and V_{in} behaves in the way expected from a simple circuit point of view.

4.7 Good thermal design

The design fault illustrated in the previous section by Fig. 4.9 may be ameliorated by changing the sign of the thermal feedback. In the very simple circuit of Fig. 4.8, this may be done in two ways: either the physical positions of Q_1 and Q_2 should be interchanged, or the d.c. level at the collector of Q_4 should be set at $3V_+/4$ so that, as equation (4.6) shows, the power dissipation in Q_4 falls with increasing V_{C4} instead of rising.

Fig. 4.9.

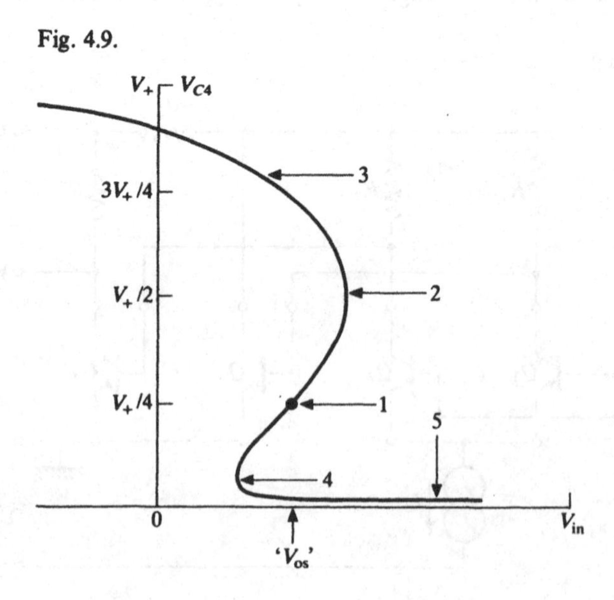

Neither of these solutions is really the answer to the problem, however. In any case we should normally be concerned with a design far more complicated than the simple four transistor circuit of Fig. 4.8, and then the power dissipation in all regions of the circuit would have to be considered, particularly in the output stages where the power dissipated would depend upon the load.

An example of this is shown in Fig. 4.10, where the output voltage–input voltage characteristic of the same 741 device used to obtain Fig. 4.7 is now shown when the external load is reduced from 30 k to only 2 k. The characteristic is quite different at small output levels. Obviously the thermal feedback from the output stage, in this particular device, is of the opposite sign to the thermal feedback coming from the intermediate stages. At high output levels, however, the power dissipation in the output transistors is very small and the characteristic shown in Fig. 4.10 takes on the same form as that shown in Fig. 4.7. Solomon [7] discusses characteristics of this complex kind.

The only way to eliminate these thermal feedback problems is to redesign the thermally sensitive input transistors, Q_1 and Q_2 in Fig. 4.8, so that they will be insensitive to changes in the temperature of neighbouring areas of the silicon integrated circuit. For example, Q_1 and Q_2 could be made up from four, or even six, isolated transistors, laid out symmetrically and cross-connected in parallel so that they would, collectively, become very insensitive to thermal gradients which developed in any direction across the integrated circuit in which they were used. This approach has been

Fig. 4.10.

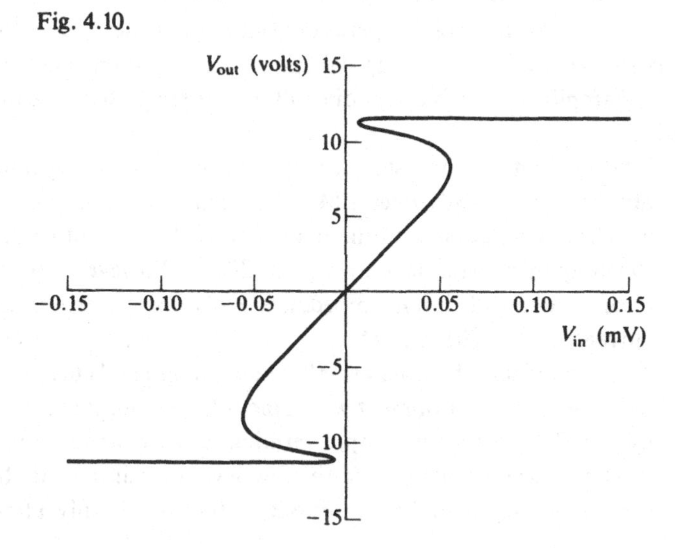

discussed by Schade [8], who gives a number of clear photographs of devices, both bipolar and MOS, in which this solution to the thermal layout problem has been used. Cross-connected arrays of transistors are used for the input circuits of all precision operational amplifiers: for example, the OP–05, which has the characteristic shown in Fig. 4.6. No change to speak of is observed in this characteristic when the load on the device is reduced.

One final problem must be considered before we leave this very interesting aspect of electronic circuit design, and that is to ask how it is possible for our closed-loop measurement and control system, shown in Fig. 4.5, to be stable if the DUT really does have a gain of the wrong sign. The answer to this question is that the negative gain shown in Fig. 4.7 is a feature of a strictly static or equilibrium characteristic. Any increase in V_{in} will, in fact, result in an increase in V_{out} but this will be followed by a relaxation back to a lower value of V_{out}, the value corresponding to the characteristic shown in Fig. 4.7. This relaxation involves a definite time delay, not a simple phase lag as would be the case if the problem involved an ordinary differential equation. These thermal problems are governed by the diffusion equation,

$$(k/\rho c)\nabla^2 T = \partial T/\partial t \tag{4.8}$$

which is a partial differential equation and implies a finite, and rather low, propagation velocity for thermal fluctuations, depending upon the thermal conductivity, k, density, ρ, and specific heat, c, of silicon [9].

4.8 Measurement of input impedance

As a final example of an operational amplifier system, let us look at a measurement system for checking the small signal input impedance of an operational amplifier. This is a very difficult measurement to make for two reasons:

(*a*) The differential, small signal, input impedance of an operational amplifier is typically over 1 MΩ. The small signal input voltage which may be applied to the input terminals, without saturating the amplifier, will be only about 20 μV, however, so that a measurement of input impedance calls for measurement of currents at the 20 pA level.

(*b*) Measurement of these very small input voltages and currents must be made in an environment which includes the output terminal of the amplifier, and this output terminal must change its voltage level by several volts during the low level measurements. In the case of an integrated circuit this output terminal is only a fraction of a millimetre away from the input terminals, and stray capacitance, or finite leakage resistance, will cause problems.

Fig. 4.11 shows a proposal for a complete operational amplifier test set which repeats what has been shown previously in Fig. 4.5 and adds the provision for measurement of I_{in}. I_{in} is measured by means of the precision operational amplifier, A_2, in Fig. 4.11, and the measurement technique would be to drive the 'control V_{out}' input at a fairly low frequency and then view the outputs of A_1 and A_2 on an XY oscilloscope, the 'Y' amplifier being a.c. coupled because we wish to see a change of, perhaps, 100 pA superimposed upon the standing input bias current of the DUT.

This display of I_{in} against V_{in} under small signal conditions will give a measure of $Z_{in} = \Delta V_{in}/\Delta I_{in}$, subject to the scaling factors set by R_2/R_3 and by R_8. However, operational amplifiers which have internal frequency compensation may show an unexpectedly low value of differential input impedance. This will be resistive and is a feature of the experimental difficulty discussed above under (*b*).

An operational amplifier with internal compensation has a gain, $G(j\omega)$, which is of the form

$$G(j\omega) = G_0/[1 + j(\omega/\omega_c)] \tag{4.9}$$

where G_0 is the very high gain we measure as a result of static tests, of the kind shown in Fig. 4.6, and ω_c is the angular frequency at which the gain has fallen by 3 db because of the effect of the internal compensation capacitor. Usually, ω_c is below $2\pi \times 10$ Hz.

Now consider such an amplifier as the DUT in the measurement set-up shown in Fig. 4.11. From the point of view of amplifier A_2, the situation is

Fig. 4.11.

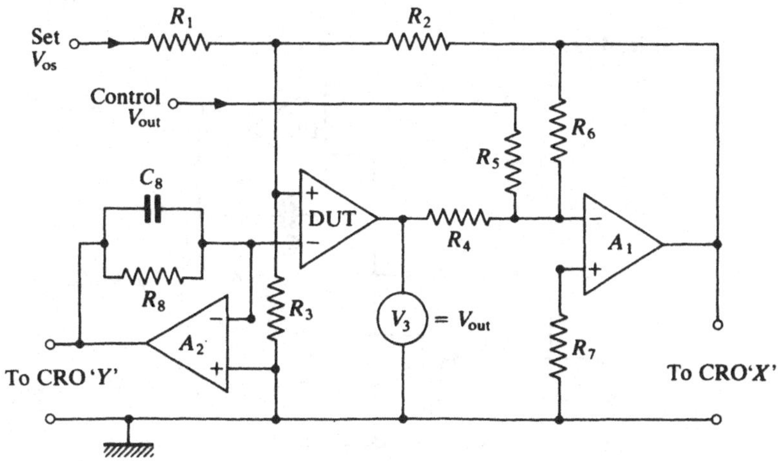

as shown in Fig. 4.12, where we show the unavoidable stray capacitance across the device, from output to inverting input, as C_s.

Now we shall have to choose a measurement frequency well above ω_c in order to measure $\Delta V_{in}/\Delta I_{in}$, simply because ΔV_{in} would otherwise have to be kept so small that ΔI_{in} would be well below the noise level of the system. This means that equation (4.9) may be reduced to $G(j\omega) = -jG_0(\omega_c/\omega)$, and Fig. 4.12 already shows this simplification in the expression for the output voltage of the DUT.

Fig. 4.12 also shows the current which is fed back to the virtual earth of A_2 from the output of the DUT via C_s. Note that this current is independent of ω. When we substitute the expression for V_{out}, the current i, in Fig. 4.12, is simply

$$i = \omega_c G_0 C_s V_{in} \tag{4.10}$$

and is quite indistinguishable from the current V_{in}/R_{in} which is flowing into the virtual earth of A_2 via the input impedance of the DUT. It follows that we see, from equation (4.10), a resistance

$$R'_{in} = 1/\omega_c G_0 C_s \tag{4.11}$$

in parallel with the true input resistance of the DUT.

Now $\omega_c G_0$ is the gain–bandwidth product of the DUT and would be typically $2\pi \times 10^6$. It follows that a stray capacitance of only 1 pF across the DUT will produce a value of $R'_{in} = 159$ k, which could be a whole order of magnitude smaller than the expected R_{in} of the DUT. The only way to solve this difficulty is to try to reduce the stray capacitance down to the level of the intrinsic stray capacitance of the DUT itself, and this is why Fig. 4.11 has been chosen as the experimental circuit for this chapter. In the Appendix, full details of the circuit and suggestions for good layout are

Fig. 4.12.

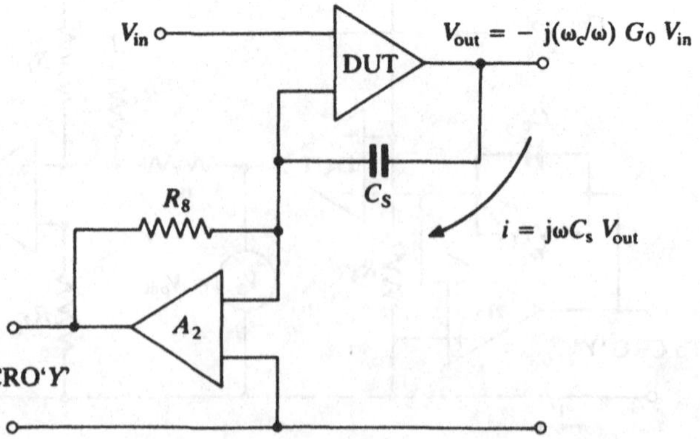

given. This measurement system is an excellent experimental exercise for getting a good feeling for layout problems of this kind.

4.9 Conclusions

In this chapter on operational amplifier systems we have looked at a few very simple circuits which are made up by treating the operational amplifier in very much the same way as we treated the transistor in the earlier chapters. Both devices have only three terminals, when looked at from an essentially circuit point of view, and both devices have extremely simple properties, when we are looking at circuits in the design stage: that is, sketching things out and looking for new ideas.

An operational amplifier system of the simple kind shown in Fig. 4.11 does, in fact, represent a very significant step forward in technique when we consider the difficulty of realising such a system before the integrated circuit became available. A textbook of 1973 [10] shows clearly how the new technique influenced analog electronics and set the scene in which quite complicated analog signal processing and active filter systems could be built up very easily, involving academically interesting concepts such as negative impedance convertors, gyrators and so on. This side of the subject is well covered by the existing texts [11].

In this chapter we have concentrated upon measurement systems intended for particular, and very specialised, measurement problems. These measurement problems have been taken from the subject of the chapter itself: the operational amplifier. This has been done to illustrate what should be a continuing interest for the experimentalist in electronics: the measurement and understanding of the detailed properties of new devices as these become available. For example, at the end of Chapter 3, some of the interesting circuit techniques which are used for input bias current cancellation were discussed. The small signal differential input resistance for many of these designs is claimed to be very high in some manufacturer's data sheets [12] and it would be very interesting to actually measure this. However, Section 4.8 has shown up some of the extreme difficulties such a measurement may involve and there is considerable scope for development in this area.

Notes

1 See Chapter 3, note 2, and also note: *Analog Integrated Circuits*, by M. Herpy, John Wiley, 1980, which is very good on circuit and system theory, and *Operational Amplifiers*, by J. Dostal, Elsevier, 1981, which is particularly good on noise problems and the measurement problems considered in this chapter. Earlier books by G.B. Clayton are all excellent and deal with the simpler problems: *Experiments with Operational*

Amplifiers, Macmillan, 1975; *Linear Integrated Circuit Applications*, Macmillan, 1976; and *Operational Amplifiers*, Newnes–Butterworth (2nd edition), 1979. There are also very valuable handbooks which originate from operational amplifier manufacturers: *Operational Amplifiers*, edited by J.G. Graeme, G.E. Tobey and L.P. Huelsman, McGraw Hill, 1981, and *Applications: Linear Integrated Circuits and MOSFETs*, RCA Corp., ref. SSD245, 1983.

2 For example, see the data sheet of the PMI dual operational amplifier OP–10. This has an application note on the three amplifier circuit as an appendix. Precision Monolithics Inc. Data Book, 1984, *Linear and Conversion Products*, pp. 5-74 to 5-85.

3 See note 1 above.

4 The strange behaviour of the 741 seems to have been first published by J. Mulvey and J. Millay, *IEEE Spectrum*, **11**, No. 9, 53–8, 1974. The experimental arrangement shown in Fig. 4.5 is based upon their ideas.

5 Despite a very early paper, 'Thermal feedback in integrated circuits', by H. Schmidt, *Phillips Electronic Applications Bulletin*, **28**, 29–39, January 1968, very little was published on this problem until the excellent tutorial paper appeared in *IEEE J. Solid State Circuits*, **SC–9**, 314–32, 1974, by J.E. Solomon. A recent textbook which corrects the past neglect of thermal design is *Analysis and Design of Analogue Integrated Circuits*, by P.R. Gray and R.G. Meyer, John Wiley, 1984 (2nd edition).

6 See Chapter 1, note 11.

7 See note 5 above.

8 O.H. Schade, *RCA Review*, **37**, 404–24, 1976.

9 Equation (4.8) is discussed in detail in *Partial Differential Equations in Physics*, by A. Sommerfeld, Academic Press, 1972, p. 34, and in *Lectures in Physics*, by R.P. Feynman, R.B. Leighton and M. Sands, Vol. 2, p. 3-6 *et seq*., Addison Wesley, 1964.

10 J.G. Graeme, *Applications of Operational Amplifiers*, McGraw-Hill, 1973.

11 See note 1 above.

12 RCA make no claim for high input resistance for the CA3193 and CA3493 (see note 19 of Chapter 3) as the data sheet only lists measurable parameters. PMI give a value of 80 MΩ for the OP–05A and OP–07A (see note 13 of Chapter 3) stating this is not measured but 'guaranteed by design'.

5

A photodiode amplifier

5.1 Photodiodes

The circuit design problem that will be considered in this chapter is very similar to the tuned amplifier problem considered in Chapter 2. In Chapter 2 we looked at the kind of circuit that would be found as an input circuit to a sensitive radio or television receiver, or in any scientific instrument where the input signal is some kind of radio frequency carrier of information.

In this chapter, a very similar kind of problem is considered but the input signal is now an optical input. The most familiar example of such a circuit is the photodiode–amplifier combination which would be found at the receiving end of an optical fibre communication link [1], although that particular problem is more specialised than the one considered here in that optical fibres usually carry digital data and the receiving circuits may be made deliberately non-linear with advantage. Optical fibres also operate at very high data rates, typically well over 100 Mbits/s, which cannot be dealt with easily in a simple laboratory.

Let us first consider the photodiode itself, one of the simple pn, or pin, variety. Connected as a linear photodetector, this device is used in the reversed biassed mode, and a typical pin silicon device would have a set of voltage–current characteristics of the kind shown in Fig. 5.1 [2].

It is clear from Fig. 5.1 that the silicon pin photodiode is a very high impedance device when it is used in the normal mode: reversed biassed. In fact, it may be represented, as a first approximation, by a current generator which produces a current linearly dependent upon the optical power input. Before making this rather drastic simplification, however, it is important to note that the photodiode has an intrinsic power gain: an increase in the optical power input may produce a larger increase in the electrical power dissipated in both the photodiode and any external load [3]. For example, consider the photodiode, with the characteristics shown in Fig. 5.1, connected in series with a 500 k load resistor to a voltage source of – 25 V. The situation would then be as shown in Fig. 5.2. For low frequency changes in radiant power input the diode would always be found working at some point along the load line *A–B*.

Fig. 5.2 shows that if the radiant power input was increased from zero to 50 μW, the diode would operate at point C where the power dissipation in the diode is 18 μA × 16 V = 288 μW, nearly six times the radiant power input, while the power dissipated in the 500 k load resistor would be 162 μW. The power dissipated in the load resistor does not increase linearly with radiant power input: it is the voltage across the load resistor which is a linear function of this power input.

Fig. 5.1. Note the change in voltage scale with change in sign.

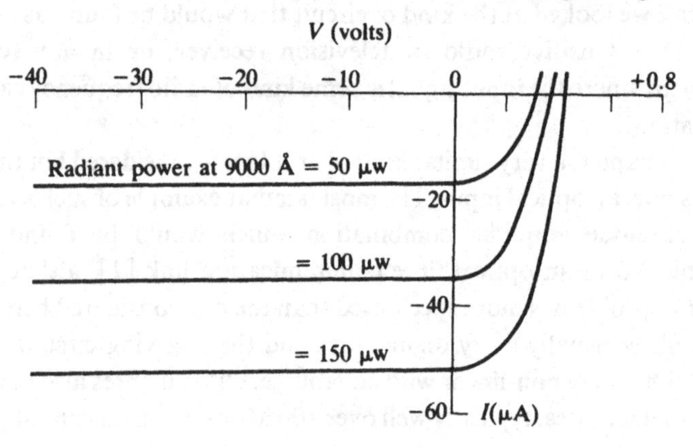

Fig. 5.2. Note the change in voltage scale with change in sign.

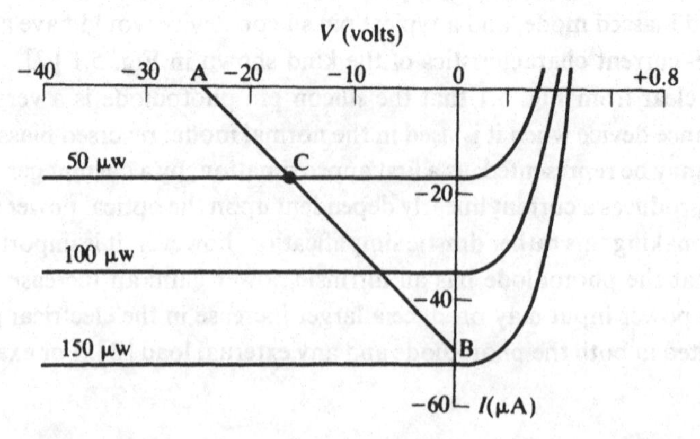

5.2 Noise and the optimum use of a signal source

This very elementary consideration of the signal source being used in this particular circuit example brings in a very important point which was tacitly ignored in Chapter 2 where the tuned amplifier was dealt with. In that chapter, all our attention was directed towards the problem of circuit stability. The problem we should like to consider now is how to deal with a real signal source, in this case the photodiode, so that the overall circuit is optimised from the point of view of sensitivity. This brings in the question of noise.

From elementary considerations we know that a signal source should be matched to the input impedance of the amplifier it is used with. This means that the internal impedance of the source should be equal to the complex conjugate of the input impedance of the amplifier. This simple kind of matching problem is the one which turns up in the tuned amplifier circuits considered in Chapter 2. For the photodiode, matters are more complicated. We are dealing with a very high impedance source in the first place; secondly, we should expect it to have some additional noise problems because of the power gain which was discussed above, and, thirdly, we are not necessarily dealing with a narrow band of signal frequencies: we would like to have a light sensitive photodiode–amplifier system which responds from d.c. up to a fairly high frequency.

To come up against the practical details at once, let us consider the simplest possible circuit shape which might be suggested as a photodiode–amplifier combination. This is shown in Fig. 5.3. The photodiode is reverse biassed by means of the easily available positive rail voltage, V_+, and its output current is fed directly into the virtual earth of an operational amplifier connected as a current to voltage converter by means

Fig. 5.3.

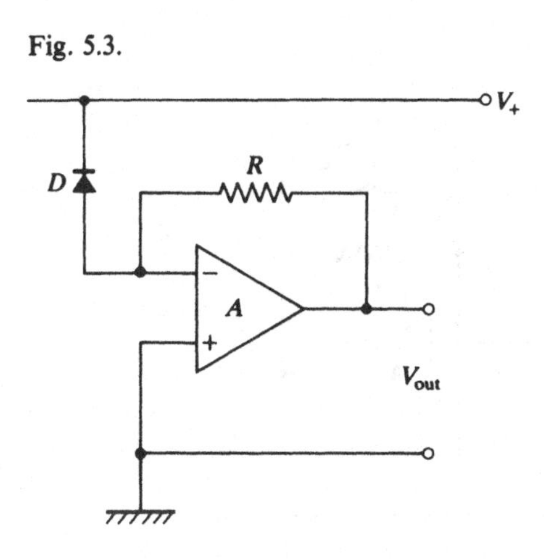

of the feedback resistor, R. The output voltage of this circuit is then simply $R \times I_R$, where I_R is the diode current, shown in Fig. 5.1, caused by the radiant power input.

Fig. 5.4 shows the equivalent circuit of Fig. 5.3, and this enables us to calculate the signal to noise ratio and sensitivity. In Fig. 5.4, i_s is the signal current, approximately 0.36 μA per μW of radiant power input as shown in Fig. 5.1, and i_{sn} is the shot noise current which must be superimposed upon any current flowing in the diode. This is given by $\overline{i_{sn}^2} = 2ei_sB$, where e is the electronic charge and B is the system bandwidth [4]. This tells us that, in an ideal case where the photodiode is perfect and there are no other sources of noise, we should have a signal to noise ratio $\sqrt{i_s/2eB}$, which is a photon counting level because the observable current is of the order of one electron per reciprocal of bandwidth. However, there is, of course, a finite current in the diode even when it is not illuminated, due to the thermal generation of hole–electron pairs, background radioactivity and leakage. This current is called the dark current and is a big factor in deciding the price of a photodiode: a cheap diode will have a dark current of perhaps 5 nA, a much more expensive one could have a dark current of only 150 pA. If we call the dark current I_{dc} we have

$$\overline{i_{sn}^2} = 2eI_{dc}B \tag{5.1}$$

The other source of unavoidable noise in Fig. 5.4 is the thermal noise in the resistor R. This has been represented by a current generator,

$$\overline{i_{tn}^2} = 4kTB/R \tag{5.2}$$

It follows that the output of Fig. 5.4 is going to be a signal, i_sR, and a noise voltage of total rms value $(\overline{i_{sn}^2} + \overline{i_{tn}^2})^{1/2}R$. The signal to noise ratio is

Fig. 5.4.

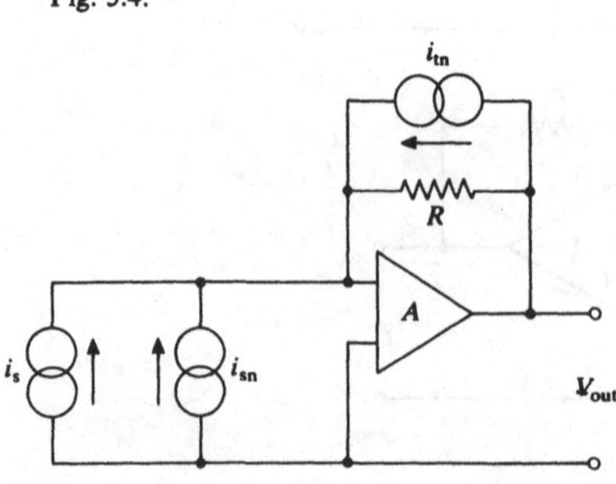

thus

$$\mathrm{SNR} = i_s/(2eI_{dc}B + 4kTB/R)^{1/2} \quad (5.3)$$

when expressed as a ratio of voltages at the output terminals of Fig. 5.4. Equation (5.3) shows that the signal to noise ratio will get better as we increase R but once we reach the point where the two terms in the denominator of equation (5.3) are equal, that is a value of R given by

$$R_{max} = 2kT/eI_{dc} \quad (5.4)$$

any further increase in R will be of negligible help in further improvement of the SNR.

Now the value of R_{max} given by equation (5.4) is very high indeed in all practical cases because the diode dark current, I_{dc}, is so small. Even for a cheap silicon photodiode with a 5 nA dark current, equation (5.4) gives a value of $R_{max} = 10\ \mathrm{M\Omega}$.

5.3 Bandwidth considerations

The circuit shown in Fig. 5.3, which led to Fig. 5.4, at first sight seems to be a good one from the point of view of obtaining a large bandwidth. This is because the photodiode is working into a very low impedance, the virtual earth of the amplifier inverting input, so that there is no voltage change across the photodiode and, consequently, the capacitance of the photodiode is of no importance. The bandwidth of the photodiode–amplifier combination should, therefore, be just below the bandwidth of the amplifier at unity voltage gain, because the circuit has 100% voltage feedback. This can be a few MHz with easily available operational amplifiers, particularly if an amplifier with an input circuit using junction field effect transistors is chosen, like the OP–15 [5], which also has the advantage of very low input noise current.

However, if a very high value of feedback resistor, R, is used in Fig. 5.3 in an attempt to obtain the highest sensitivity and optimum signal to noise ratio, along the lines described in the previous section, any wide-band properties of the amplifier, A, will be radically reduced by the unavoidable stray capacitance, C, which must exist across R. This is not a trivial problem: for example, the overall bandwidth is not simply $1/2\pi CR$ but usually somewhat better. The reason for this is shown in Fig. 5.5, where the feedback loop of Figs. 5.3 and 5.4 has been opened and the signal sources replaced by their internal impedances. Because these internal impedances are very high we find that the input impedance of the amplifier itself plays a very important role in this problem. The input capacitance of the amplifier is increased by the photodiode capacitance, so that this does, after all, influence the performance of the circuit.

If we reduce the value of R we can begin to realise the full bandwidth potential of the devices but the noise now depends mainly upon the value of R. Some excellent designs have been published along these lines. For example, Mitchell *et al.* have published [6] a design using silicon bipolar transistors and an InGaAs photodiode which has an overall bandwidth of 260 MHz. This is built not simply as a monolithic silicon integrated circuit but as a hybrid circuit involving thin film and chip components mounted on a ceramic substrate. This kind of advanced constructional technique can reduce the shunt capacitance across the feedback resistor to 0.02 pF, even though the feedback resistor is still a fairly high value at 20 k. A circuit of similar performance, using GaAs field effect transistors and silicon bipolar devices, again built as a hybrid circuit, has been published by Ogawa and Chinnock [7].

5.4 A shape for an experimental circuit

For our experimental photodiode–amplifier we shall aim at a very modest bandwidth so that conventional constructional techniques may be used. We shall also look at an alternative to the circuit shape shown in Fig. 5.3. This alternative is shown in Fig. 5.6 and brings us back to the point which was made at the beginning of this chapter: the intrinsic power gain of the photodiode.

In Fig. 5.6 the photodiode, D, does not work into a very low impedance, as it did in Fig. 5.3, but directly into a high impedance load, R. The noise calculation and the sensitivity calculation for this circuit lead to exactly the same result, however, as the previous calculations for Fig. 5.3. In Fig. 5.3, no power gain was realised from the photodiode because it was loaded, virtually, with a short-circuit. In Fig. 5.6, we use a high value of resistive load, R, but the power delivered by the photodiode is simply dissipated in this resistor and there is no advantage in doing this. An advantage would be obtained if, instead of delivering power to a simple resistor, R, we were able to deliver this power into the input impedance of an amplifying device like a transistor. This is a possibility at very high frequencies and will be looked at briefly at the end of this chapter.

Fig. 5.5.

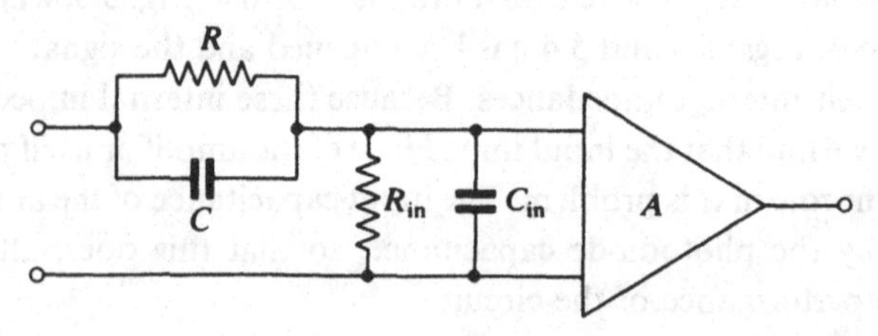

The advantage of using a circuit shape of the kind shown in Fig. 5.6 is that it leads us to consider a different way of dealing with the problem of obtaining a large bandwidth and high sensitivity, when the constructional technique we have to use makes it impossible to reduce circuit stray capacitance very much. In Fig. 5.6, the photodiode capacitance, which will be at least 2 pF just for the junction and at least a further 2 pF for the connection between the junction and the rest of the circuit, is directly in parallel with the load resistor, R. The load resistor will have some self capacitance as well, but the amplifier, A, is now a simple voltage follower and its input capacitance will be really negligible. It is possible, however, that there will be quite a large capacitance to ground at the amplifier input. If all these separate capacitances add up to some total, C, then the circuit shown in Fig. 5.6 will have an overall bandwidth $(1/2\pi CR)$, always assuming that the voltage follower, A, has a wider bandwidth. From this initial point of view, Fig. 5.6 is inferior to Fig. 5.3.

The effect of the unwanted capacitance, C, can be reduced very considerably, however. This is done by a technique which has become known as bootstrapping [8]. The idea is to use feedback so that the voltage across some unwanted capacitance is forced to remain constant and, since $i = C\mathrm{d}V/\mathrm{d}t$, making $\mathrm{d}V/\mathrm{d}t$ equal to zero is the same thing as making C equal to zero.

Fig. 5.7 shows a simple modification which will make the circuit shown in Fig. 5.6 practically independent of photodiode capacitance. The modification is to add C_2R_2, ensuring that $(1/2\pi C_2R_2)$ is much smaller than $(1/2\pi CR)$. Positive feedback, via C_2, then makes the voltage across the diode, D, *constant* at the higher frequencies, regardless of the light signal input. This removes the effect of the photodiode capacitance but leaves us with the problem of any stray capacitance between the positive input terminal of the amplifier, A, and ground.

Fig. 5.6.

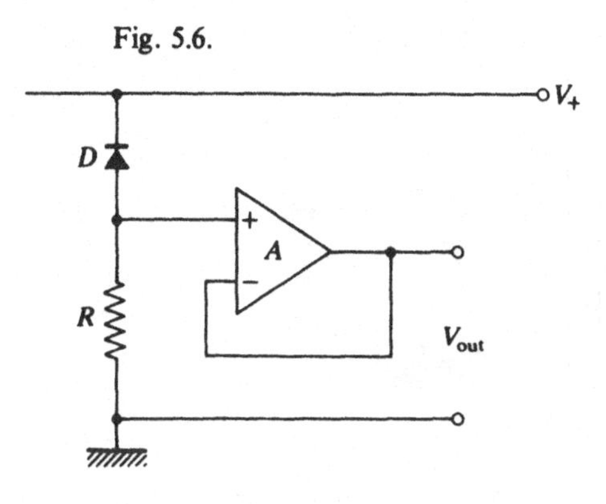

The bootstrapping idea can be extended to cover far more than just the diode capacitance, however, and to see an example of this we go directly to the final circuit which is proposed for experimental work.

5.5 Details of an experimental circuit

The circuit which has been chosen for experimental work in this chapter is shown in Fig. 5.8. Note first of all that the 3100 wide-band operational amplifier [9] and the 2N2222A transistor form a simple output amplifier which has a voltage gain of 100, a bandwidth of 1 MHz and is able to drive a terminated 50 Ω cable which may then be used to connect the circuit to an oscilloscope or other test equipment.

The part of Fig. 5.8 which corresponds to Fig. 5.7 involves the 40841 MOST, a dual gate device [10] connected as a source follower. This corresponds to the voltage follower, A, in Fig. 5.7. Bootstrapping of the photodiode, D, in Fig. 5.8 is accomplished via C_2R_2, as in Fig. 5.7, but this is not the only part of this circuit which uses the bootstrapping idea. Gate 2 of the 40841 is also bootstrapped to the output of the source follower via C_3 and this makes the true input capacitance of this source follower very small indeed. If a single gate device had been used, the gate to drain capacitance would have remained in parallel with the input capacitance and would have degraded the high frequency performance.

The photodiode load in Fig. 5.8 is R_1. To be exact, it is R_1 plus the parallel combination of R_3 and R_4, but R_3 is made about 1 % of R_1 and is only there to provide a negative bias for the 40841, and is set to such a value that the output of the circuit is zero for zero light input. The bootstrapping precautions make it possible to reduce the capacitance across R_1 to simply

Fig. 5.7.

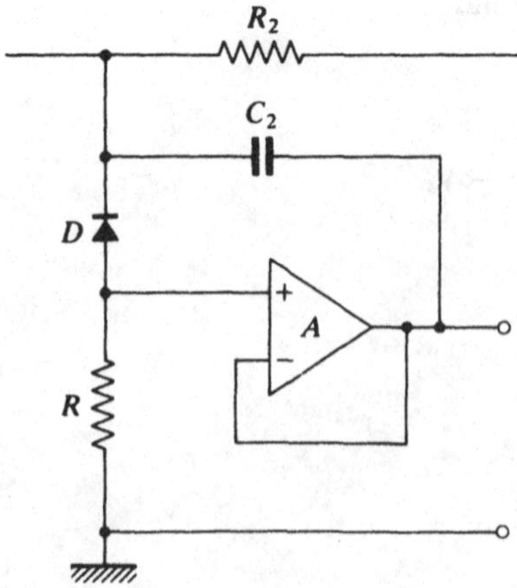

the self-capacitance of this resistor itself. With very small resistors of the metal film type [11], this self-capacitance is of the order of only 1 pF so that this means, if we arbitrarily choose to make our system bandwidth 1 MHz, we shall not be able to use a value of R_1 greater than 150 k. Now this is well below the maximum value suggested by equation (5.4) and means that the noise from the photodiode load resistor, R_1, will be the dominant noise source in this circuit. The photodiode noise, equation (5.1), is quite negligible in comparison. Nevertheless, the noise from R_1 is only about 40 μV at the output of the 40841 and this is why the simple amplifier with a gain of 100 is added. The noise from the output socket of the circuit shown in Fig. 5.8 should be about 4 mV rms, and this is an ideal level for working into any general purpose oscilloscope.

The overall performance of the circuit shown in Fig. 5.8 can be assessed by using a light emitting diode [12] and a pulse generator as a signal input. Depending upon the choice of photodiode, some rather unexpected behaviour may be observed on a long time scale [13], but the important parameter to measure is the rise time of the photodiode–amplifier combination. With two 1 MHz bandwidth circuits in cascade: the 40841 with R_1 at 150 k, and the self-capacitance of R_1, and then the 3100, with a gain of 100, the rise time of the whole system should be less than 0.4 μs. It is very easy to verify that this depends upon the capacitance in shunt with R_1

Fig. 5.8.

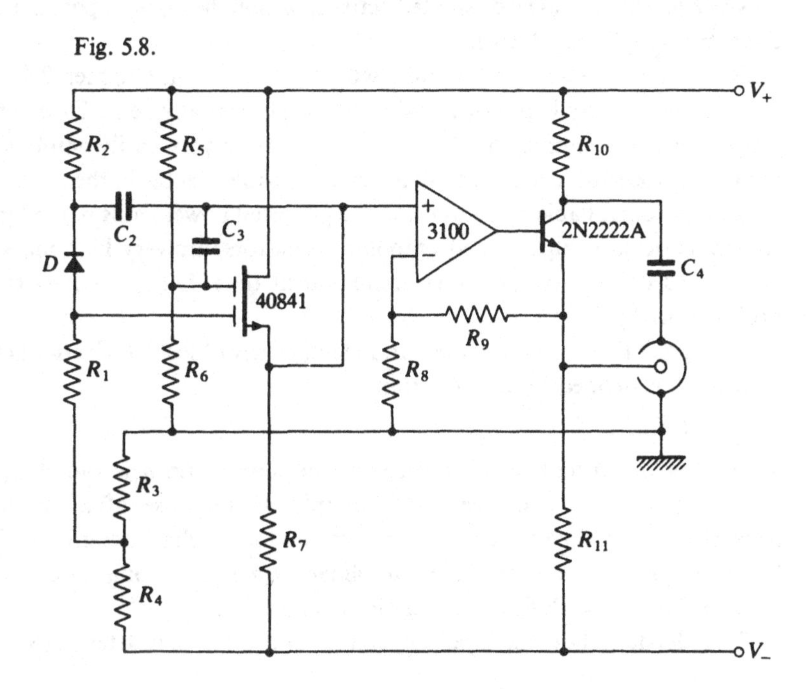

by adding a very small capacitance in parallel with R_1. Similarly, it is easy to verify that the photodiode capacitance does not degrade the performance. This is done by adding a very small capacitance in parallel with the diode, D, in Fig. 5.8. Full constructional details for Fig. 5.8 are given in the Appendix, and it is a particularly valuable circuit to experiment with because of the need to consider the layout carefully so that the bootstrapping really does reduce the stray capacitance across R_1.

5.6 A general discussion of power gain

To conclude this chapter it is interesting to look at a general problem in electronics which has been brought up in the discussion of the photodiode amplifier. This is the question of designing a circuit to give power gain, an essential feature of any circuit or system which plays a part in the transmission of information.

It was shown, in Section 5.1, that the photodiode itself possessed intrinsic power gain and that it could deliver power to an external resistor, R. In all the circuits which we looked at, however, this external load resistor was either made effectively zero, as in Fig. 5.3, or it was a simple passive resistor, as in Fig. 5.6 and in our final experimental circuit, Fig. 5.8. In other words, nothing useful is done with this available power whereas it might be expected that a more sensible circuit design would ensure that the small power available would be dissipated actually within the input impedance of the active amplifying device.

This brings us back to a remark which was made in Chapter 2 [14] concerning the very high power gains which can actually be realised with simple electronic amplifying devices but cannot be very easily employed because of the problem of making the circuit usable: usable in the sense of stable and controllable performance. The potential power gains of modern electronic devices, bipolar and unipolar transistors, are very high indeed and this is because of their very small transit time. Let us look at this problem briefly.

The transit time, τ_t, of a device, or, in fact, of any object which conducts electricity, is defined by the equation

$$I = q/\tau_t \tag{5.5}$$

where I is the current passing through the device and q is the charge contained within the device. For example, in the case of a bipolar transistor, I would be the collector current and q the charge contained in the base. The device works as an amplifier by means of some externally applied signal modulating the magnitude of q [15].

To understand how the input impedance of the device is determined by the transit time we need to consider a particular kind of device, and the simplest one to look at is the unipolar, or field effect, transistor because this has a purely capacitive input impedance at zero frequency. A change in voltage, v, on the gate of the FET will then produce a change of charge, q, contained in the channel, according to the simple equation $q = C_{in}v$. However, at any finite frequency the *mobile* charge in the channel, which can only come from the source, takes a finite time to become established. In the first approximation we can write

$$\tilde{q} = C_{in}\tilde{v}/(1 + s\tau_t) \tag{5.6}$$

where s is the Laplace transform variable and $\tilde{q}$ and $\tilde{v}$ represent the Laplace transforms of the functions $q(t)$ and $v(t)$.

From this point of view, the operational input admittance of the device is $s\tilde{q}/\tilde{v}$ or

$$Y_{in}(s) = sC_{in}/(1 + s\tau_t) \tag{5.7}$$

and to see what this means we go to a single frequency representation by writing $j\omega$ for s, in equation (5.7), and separating the real and imaginary parts:

$$Y_{in}(j\omega) = \frac{j\omega C_{in}}{(1 + \omega^2\tau_t^2)} + \frac{\omega^2 C_{in}\tau_t}{(1 + \omega^2\tau_t^2)} \tag{5.8}$$

Equation (5.8) shows that the input admittance of a power amplifying device does have a positive real part and that this increases rapidly with frequency until we begin to approach the really high frequency range where $\omega\tau_t > 1$. Fig. 5.9 shows some experimental data for a high frequency MOST: the 3N204 [16], the full lines being drawn through measured values of

$$Y_{in} = g_{in} + jb_{in} \tag{5.9}$$

while the broken lines are simply extrapolations down to the lower frequencies made according to equation (5.8), assuming that $\omega\tau_t \ll 1$. It is clear that the simple model we have used to arrive at equation (5.8) is reasonably accurate and that, even at 1000 MHz, $\omega\tau_t$ is still well below unity, which means that τ_t is well below 100 ps, as we would expect from an n-channel MOST with a channel length of perhaps only a few microns, operating at the saturation velocity for electrons in silicon of 10^5 m/s [17].

The important fact to be seen from Fig. 5.9, from the point of view of our photodiode–amplifier design problem, is that the real part of the input impedance, for the kind of first stage device we have chosen, is well over 1 MΩ over the band of frequencies being used. In fact it is about a factor of twenty greater than the values shown in Fig. 5.9, because, in Fig. 5.8, the MOST is being used as a source follower whereas Fig. 5.9 shows data for the common source connection. It follows that negligible power is actually delivered to the first stage device: its true power gain is very high indeed. To make the circuit usable, we have to shunt the very high, and, as Fig. 5.9 shows, very frequency dependent, input impedance with a much lower, constant, impedance.

This is not the situation at all when a circuit is designed for the very high frequencies. In the circuits described by Mitchell *et al.*, and by Ogawa and Chinnock [18], for example, the input device is working at a frequency where the input impedance has a real part of perhaps only 1000 Ω. At the simplest level, we are now back with a design problem of the kind illustrated in Fig. 5.5. There, the input impedance of the amplifier was important because, at low frequencies, all the other impedances were so high. Now, we have to consider the true input impedance because, at really

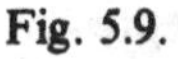

Fig. 5.9.

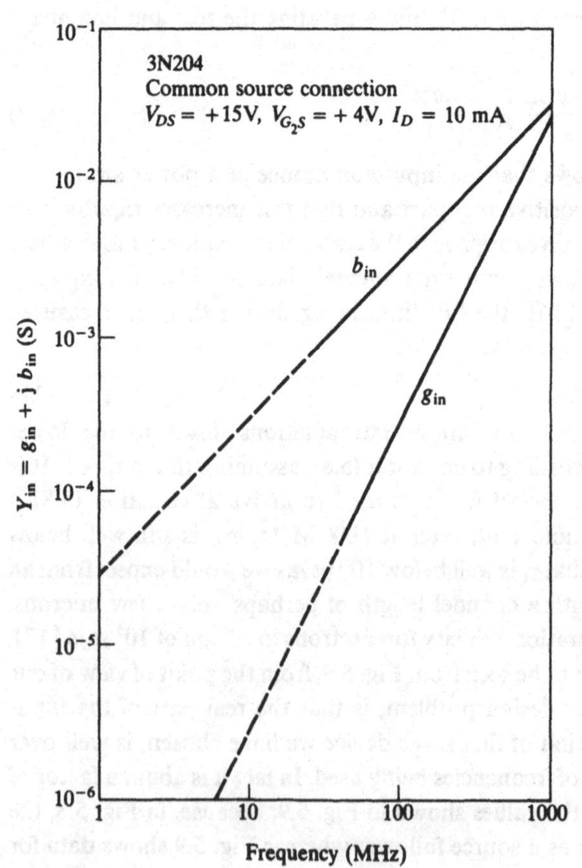

high frequencies, it becomes so low. The power gain being obtained from the first active device in the circuit is then not very much less than the maximum possible power gain. Exactly the same argument applies when the first active device is a bipolar transistor but then there is an additional real part added in parallel with the input impedance due to recombination in the base.

Notes

1 An excellent text on fibre-optic communications is the one edited by C.P. Sandbank: *Optical Fibre Communication Systems*, John Wiley, 1980. This book is particularly interesting for its detailed photographs of hardware.
2 The characteristics shown are for the Hewlett-Packard 5082–4200 series.
3 The physics of this power gain is essentially that a single photon having, in the case of silicon, an energy of about 1 eV, will produce an electron–hole pair in the electric field of the photodiode junction. The reverse voltage across this junction is normally far greater than one volt.
4 *Low Noise Electronic Design* by C.D. Motchenbacher and F.C. Fitchen, John Wiley, 1973, is a very useful text on noise theory and practice. A good introduction to the topic is Chapter 7 of *The Art of Electronics* by P. Horowitz and W. Hill, Cambridge University Press, 1980.
5 Precision Monolithics Inc. Data Book, 1984, p. 5-90.
6 A.F. Mitchell, M.J. O'Mahony and B.A. Boxall, *Electronic Letters*, **19**, 446–7, 1983.
7 K. Ogawa and E.L. Chinnock, *Electronic Letters*, **15**, 650–2, 1979.
8 F.C. Williams gives some brief remarks on the origin of this term in electronics in his paper, 'Introduction to circuit techniques for radiolocation', *J. IEE*, **93**, Part IIIA, No. 1, 289–308, 1946.
9 See note 1 of Chapter 3.
10 See note 7 of Chapter 2.
11 For example, the SFR25 type from Mullard.
12 The LED must have a fast rise time. For example, the Hewlett-Packard 5082–4880 series, which have a 15 ns turn-on time.
13 See pp. 192–5 of Sandbank, note 1 above.
14 See note 9 of Chapter 2.
15 An excellent tutorial paper on this way of looking at amplifying devices has been published by R.D. Middlebrook, *Proc. IEE*, **106B**, Suppl. 15–18, 887–902, 1959. This puts things in a good historical perspective.
16 RCA Data Bulletin File No. 959.
17 R.S.C. Cobbold, *Theory and Applications of Field Effect Transistors*, Wiley-Interscience, 1970, p. 104, Fig. 3.19(*b*).
18 Notes 6 and 7 above.

6

Digital circuits

6.1 Switches

The circuits used in modern digital systems are the simplest of all non-linear electronic circuits because the active, or power amplifying, devices are used only as switches. It is important to realise this is only true for the most recent times [1]. In an interesting review article by Renwick [2], written in 1960 at just about the time that semiconductor devices were taking over, it is made clear that a number of very complicated techniques had to be employed in computers at that time, in order to reduce the quantity of hardware involved. Perhaps the most surprising feature of these early techniques, looked at from today, was that direct coupling between circuits could not be used. Digital signals had to be carried by resistance–capacitor networks and d.c. restoration clamps had to be introduced repeatedly.

However, this was not the case in the electric computers of the 1930s, which used relays in both arithmetic–logic unit and memory [3]. To begin our discussion, it is interesting to look briefly at the way logic is done with relays because the electromagnetic relay does, in fact, have certain ideal properties which are not found in electronic switches. By reminding ourselves of these problems, which have had to be solved in the course of computer development, many of the particular features of the circuits which we use today, and will use in the future, become more understandable.

6.2 Logic gates

Fig. 6.1 shows a simple three input logic gate realised using relays. In this circuit a contact closes when its relay coil is energised. In Fig. 6.1 the *D* contact will only close if *A*, *B* and *C* are all energised so that if the closed state is represented by the binary '1' and the open state by '0' we have a logical relationship

$$A \cdot B \cdot C = D \tag{6.1}$$

between inputs and output, and would call Fig. 6.1 an AND gate. However, if we choose to represent the open state as '1' in Fig. 6.1, this circuit implements the logical relationship

$$\bar{A} + \bar{B} + \bar{C} = \bar{D} \tag{6.2}$$

where we continue to give the symbols A, B, C and D the same meaning as in equation (6.1). From the point of view of equation (6.2) we now seem to have an OR gate. We have here, of course, a simple reminder of de Morgan's relationship,

$$\bar{A} + \bar{B} + \bar{C} = \overline{A \cdot B \cdot C} \tag{6.3}$$

and the fact that any primitive logic gate can be looked upon as an OR gate or as an AND gate: it depends upon our choice of reference for the '1' and '0' states.

Fig. 6.2 shows another possibility for relay logic. This uses a constant current source and has contacts in *parallel*, in contrast to Fig. 6.1 which used a voltage source and had contacts in *series*. The circuit of Fig. 6.2 implements the logic function

$$A + B + C = \bar{D} \tag{6.3}$$

when the '1' state is represented by a closed contact, and would be called a NOR gate.

Fig. 6.1.

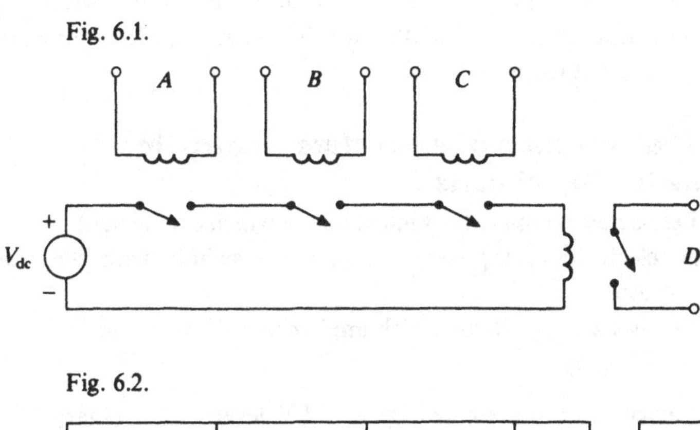

Fig. 6.2.

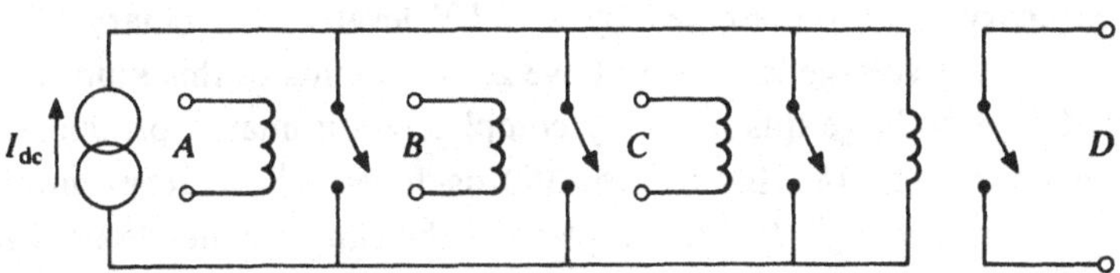

The two very primitive logic gates shown in Figs. 6.1 and 6.2 will be considered again towards the end of this chapter, when two techniques of very large scale integration (VLSI), CMOS and I^2L, are compared. For the moment, however, let us look again at the electromagnetic relay and use it to illustrate some of the essential features of an ideal digital device.

6.3 The ideal digital device

The electromagnetic relay, a device consisting of one or more energising coils operating one or more electrical contacts, does have a number of properties which make it an ideal device for building digital systems. These are worth listing because they emphasise the problems that exist with the much faster, smaller and cheaper switching devices used to build digital systems at the present time.

The remarkable features of the relay are:

(1) The complete isolation between input and output. This means that there can be any voltage difference between input and output circuits which may be convenient: input and output are floating.

(2) Multiple inputs can be easily arranged by using several coils. Isolation between inputs at low speeds is good.

(3) Multiple outputs are particularly easy to arrange and isolation between them is excellent.

(4) The power gain of the device is very high indeed: it is simply the product of the hold-off voltage and the rated contact current, several watts, divided by the power input needed to operate the relay, a few milliwatts. A related feature is that the power loss in the output circuit, or, in other words, the voltage drop across the closed contact, is negligible.

6.4 The evolution of solid state electronic switching circuits

Let us now compare the kind of device which can be built up from solid state electronic components, to make a switch, with the relay discussed above.

Fig 6.3 shows a logic circuit which implements the function

$$A \cdot B \cdot C = \bar{D} \tag{6.4}$$

when a binary '1' is represented by a +3 V level and a binary '0' is represented by a voltage level just above zero. Circuits of this kind were produced in very large quantities by completely automated production techniques in 1964 for the IBM System/360 machines [4] and are examples of diode–transistor logic (DTL). In place of the simple contacts used in relay logic, we find diodes being used as switches and the on or off condition is decided by changing the voltage level across the diode. Unlike a relay contact, there is quite a large voltage difference across a diode when it is conducting and this, along with the power loss it involves, has to be corrected by means of a simple transistor amplifier at the output end of the diode switches before we can continue to build up a system by cascading these circuits. The input and output voltage levels must be compatible.

Fig. 6.3, although it was manufactured by a completely automated technique, was not a silicon integrated circuit. It involved discrete diodes and transistors and thick film resistors. When we consider manufacturing Fig. 6.3 as a monolithic silicon integrated circuit, there is one feature which stands out at once and that is that D_1, D_2, D_3 and D_4 all have a p-type silicon region, and all these regions are connected together. Why not make this simply *one* p-type region in the integrated circuit? The fifth diode, D_5, has to remain separate, but this diode is only needed to ensure that the base current of Q_1 can really be reversed for a short time when input A, or B, or C, goes to the low voltage level. When we think more deeply about this we can see that D_5 can be dispensed with, along with the need for a negative supply. This will become clear in the next section.

6.5 Transistor–transistor logic (TTL)

Fig. 6.4 shows one of the TTL circuits which evolved from Fig. 6.3. It was realised that the diodes D_1 to D_4, in Fig. 6.3, could be replaced by an npn transistor having as many *emitters* as the number of circuit inputs required. In the simple integrated circuit process of the sixties this was a

Fig. 6.3.

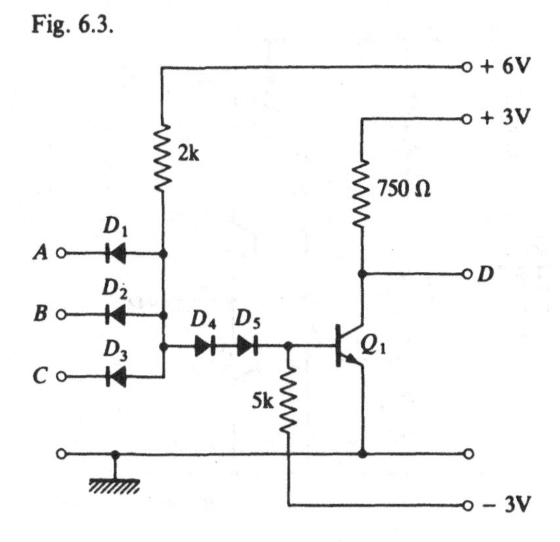

natural development: the emitters were formed by the last diffusion step [5]. By using a transistor, Q_1 in Fig. 6.4, instead of the diodes of Fig. 6.3, we can reverse the base current of Q_2, and so turn Q_2 off quickly, far more effectively because this reverse base current is the collector current of Q_1 as it is turned on. When the inputs A, B and C are high in Fig. 6.4, the collector junction of Q_1 is simply forward biassed, holding Q_2 on in very much the same way as Q_1 was held on in Fig. 6.3 by forward bias of D_4 and D_5. The inverted current gain of Q_1 in Fig. 6.4 is quite negligible, as it must be for any transistor made by the standard bipolar integrated circuit process.

What is particularly interesting about the early TTL circuits is the type of output circuit which evolved. The idea of using a simple transistor with a resistive load, as shown in Fig. 6.3, gets replaced in Fig. 6.4, and in the circuits we shall discuss below, by a circuit shape which is very simple indeed. This circuit shape is shown in Fig. 6.5; the high or low voltage level at the output terminal, D, is determined by closing S_1 or closing S_2. The circuit must be arranged so that it is impossible to either close or open both switches together and, of course, the switches must be realised using transistors.

An ideal solution to this problem is shown in Fig. 6.6: a circuit often referred to as a complementary emitter follower. This circuit is *not* used in any TTL integrated circuit simply because the process used to make these circuits allows the use of transistors of only one polarity. Returning to Fig. 6.4, we see that the Q_1 of Fig. 6.6 turns up as a Darlington pair [6], Q_5 and Q_3 in Fig. 6.4, while the Q_2 of Fig. 6.6 has to be an npn device, Q_4 in Fig. 6.4.

Fig. 6.4.

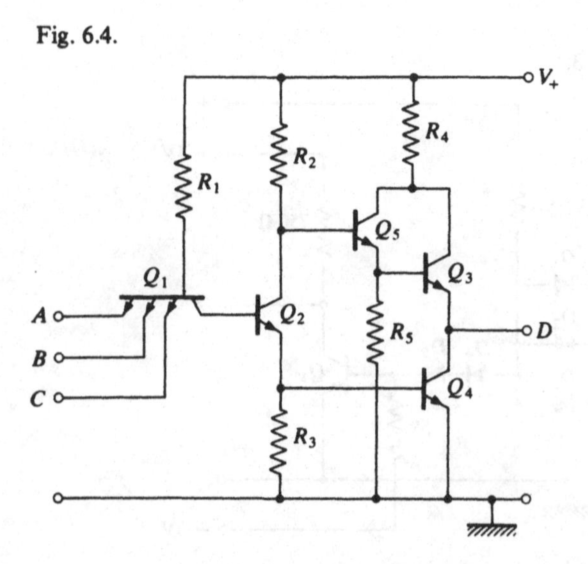

Q_4 can only be held off in Fig. 6.4 by reducing its base–emitter voltage to zero. This is easily done by turning Q_2, in Fig. 6.4, completely off: the result of having any of the inputs, A, B or C, at the low level. In this state Q_5 will saturate and Q_3, which will be kept active because Q_5 prevents forward bias of its collector junction, will be on.

The problem in the circuit of Fig. 6.4 is how to turn Q_5 and Q_3 off when Q_4 is on. It is the use of the Darlington pair which makes this possible because a Darlington pair needs about 1.4 volts across its input to be held on. The worst case in this circuit, from the point of view of keeping Q_5 and Q_3 off, is when the output, D, is low. The base of Q_5 must then be held well below 1.4 volts. This can be done by saturating Q_2, which would then have about 0.2 volts across it. The base of Q_4 will be at about 0.7 volts. It follows that the collector of Q_2, and thus the base of Q_5, will be at about 0.9 volts, well below the 1.4 volts mentioned above.

The output circuit shown in Fig. 6.4 was used in the very first silicon integrated circuits which were manufactured for logic applications [7]. A

Fig. 6.5.

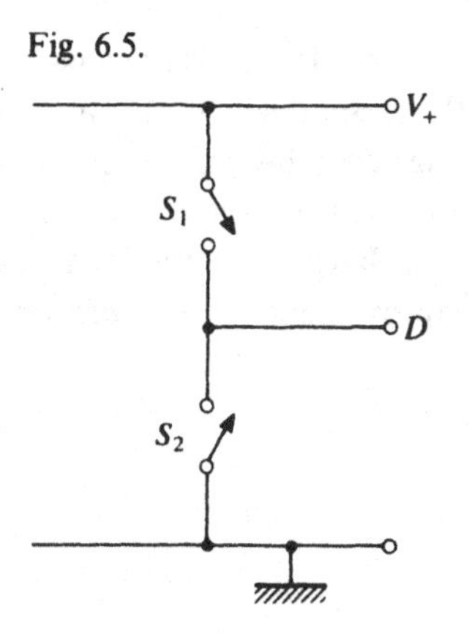

Fig. 6.6.

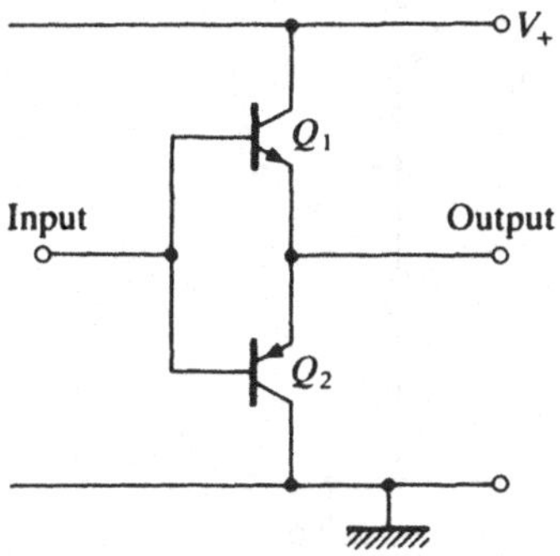

remarkable development was that another form of this circuit appeared, which is shown in Fig. 6.7. This has the same *shape* as Fig. 6.4, and the way in which Q_3 is held off appears to be the same: the diode forward voltage adds to the emitter junction forward voltage. However, the diode prevents any possibility of reversing the base current of Q_3 to give fast turn-off. In Fig. 6.4, it is R_5 which allows this. Also, it is possible for Q_3 to saturate in Fig. 6.7. In Fig. 6.4, Q_5 prevents Q_3 from saturating. Compared to Fig. 6.4, Fig. 6.7 is not a good piece of circuit design.

Finally, consider Fig. 6.8. This is one of the most familiar of all digital electronic circuits from the sixties [8] with its values: $R_1 = 4$ k, $R_2 = 1.6$ k, $R_3 = 1$ k and $R_4 = 130\,\Omega$. Realised as an integrated circuit, with about six gates of the kind shown in Fig. 6.8 on a single chip, connected into various useful configurations, this 'Standard TTL' became a major component in the electronics industry for about ten years. Fig. 6.9 shows the economic facts [9]. After an exponential looking take-off, the sales data join the cyclic and expanding form, characteristic of the economic activity of the United States at that time, and then, after about ten years, sales virtually disappear.

Fig. 6.8, despite its obvious commercial importance, is not a particularly good piece of design compared to the design which appears to have originated this TTL family: Fig. 6.4. Fig. 6.8 allows Q_3 to saturate and, while it is possible for Q_2 to draw current out of the base of Q_3 for fast turn-off, this is only possible while D_1 still has charge stored in it. At least the position of D_1 in Fig. 6.8 is a great improvement on the position of the

Fig. 6.7.

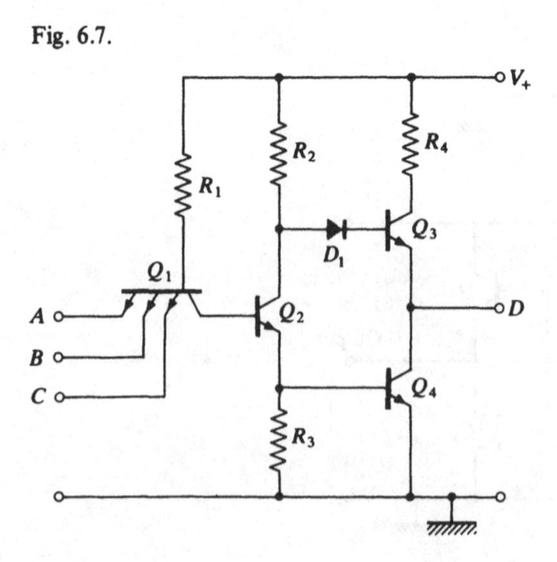

diode in Fig. 6.7. D_1 in Fig. 6.8 is taking the full emitter current of Q_3, so there should be more charge stored in D_1 than in the base of Q_3. In Fig. 6.7, D_1 only carries the base current of Q_3. Nevertheless, turn-off of Q_3 is a problem in this circuit as measurements by Tizzard and Turner [10] clearly show. If Q_2 is not saturated before Q_4, it is possible for both Q_3 and Q_4 to be on together and a large transient current will be drawn from the power supply.

6.6 Schottky TTL

The previous section has given the history of the development of Standard TTL, concentrating attention upon the interesting circuit shape which was adopted for the output of the circuit and showing how a quite good design, Fig. 6.4, could become a rather bad one, Figs. 6.7 or 6.8, by the need to simplify the circuit, and yet the end result was of unquestioned commercial success. This is really an example of the fact that people make their own circuits but they are not free to choose the parts, or more precisely, the process. The early integrated circuits of the mid-sixties had to be kept as simple as possible.

A return to the circuit shape of Fig. 6.4 was made in the early seventies when a more complex process for making bipolar digital integrated circuits came into use. This process allowed the use of so-called Schottky transistors. A Schottky transistor is a bipolar transistor which is clamped, so that it cannot saturate, by connecting a Schottky diode in parallel with its collector junction. Fig. 6.10 shows the essential details of this idea.

Fig. 6.8.

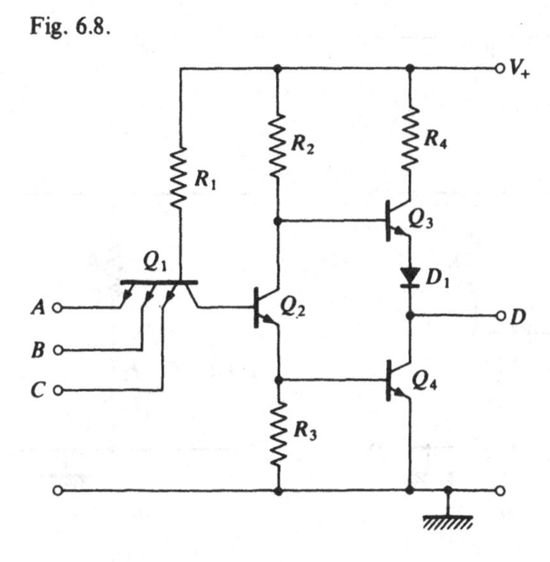

In Fig. 6.10(*a*), an ordinary discrete npn transistor, Q_1, is shown with a Schottky barrier diode, D_1, connected in parallel with its collector junction. There is a simple resistive load, *R*, from the collector to the positive supply. When an input signal turns Q_1 on, the collector voltage falls but its minimum value will no longer be $V_{CE(SAT)}$, typically 100 mV, but the higher voltage $V_{BE(SAT)} - V_D$, where V_D is the forward voltage of the Schottky diode. A good choice for V_D would be 300 mV. A silicon transistor with only 300 mV forward bias across its collector junction will still be in the active mode.

A Schottky diode [11] is a metal–semiconductor diode and its forward voltage can be determined by a suitable choice of materials. What is really vital about the Schottky diode, however, is the ease with which it can be

Fig. 6.9. Sales in the USA of standard TTL devices.

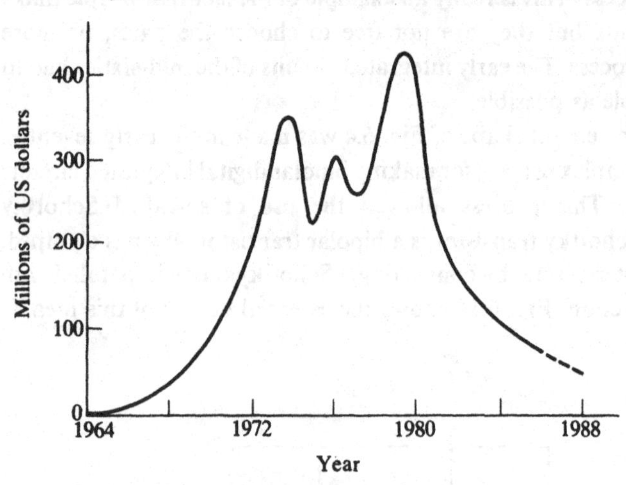

Fig. 6.10.

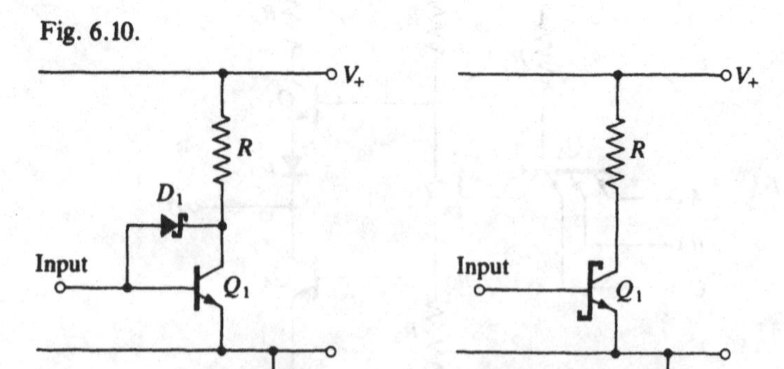

incorporated into the silicon integrated circuit process: it simply involves making the normal base metallisation contact overlap on to the n-type collector region, which may be ion-implanted to obtain the required Schottky barrier height [12]. Such a 'Schottky transistor' is represented by the symbol shown in Fig. 6.10(*b*).

The idea of clamping a transistor with a diode, as shown in Fig. 6.10(*a*), is usually attributed to Baker [13]. The reason for doing it, of course, is to improve circuit speed: the charge stored in the base of the transistor when it is on is minimised and the charge stored in the clamping diode is negligible because it is a hot carrier device.

Fig. 6.11 shows a typical circuit for Schottky TTL. It is clear that we have returned to the Darlington connection for Q_5 and Q_3, found in the earliest circuit shape proposed, Fig. 6.4. Fig. 6.11 shows a new feature, however, and that is Q_6 and R_6. In place of a simple resistor, R_3 in Figs. 6.4, 6.7 and 6.8, we now have what has become known as an 'active pull-down' [14].

Fig. 6.12 shows the results of some comparative measurements which illustrate the function of R_3, R_6 and Q_6 in Fig. 6.11. By making up this part of the integrated circuit as a discrete component circuit, using values $R_3 = 390\,\Omega$ and $R_6 = 330\,\Omega$, which are very close to the values actually used in the real circuit, and making up Q_6 by means of a general purpose npn transistor, clamped with a Schottky diode of 400 mV forward voltage, we can make a simple two terminal measurement of the current–voltage characteristic. This is shown in Fig. 6.12, where it is compared with the

Fig. 6.11.

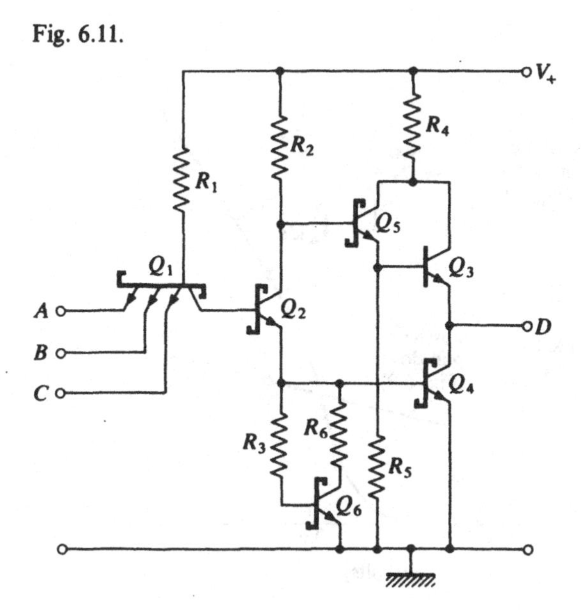

characteristic of the linear resistor, $R_3 = 1$ k, which is used in Fig. 6.8. At high values of applied voltage, that is for V well over $V_{BE} = 0.65$, we may write the current–voltage characteristic as

$$I = (V - V_{BE})/R_3 + [V - (V_{BE} - V_D)]/R_6 \tag{6.5}$$

which is a straight line extrapolating back to cut the $I = 0$ axis at a voltage

$$V' = V_{BE} - V_D R_3/(R_3 + R_6) \tag{6.6}$$

and, for $V_{BE} = 0.65$ and $V_D = 0.4$, V' comes out to be around 0.43, as shown in Fig. 6.12. Below this voltage, the active pull-down draws negligible current.

What this means is that, as Q_2 turns on in Fig. 6.11, virtually all its emitter current will flow into the base of Q_4 until nearly the full charge which is required is established. Then the current in the active pull-down rises sharply so that the final state would be close to the point P_2 in Fig. 6.12, where we have again assumed a steady state V_{BE} of 0.65 V.

In contrast, R_3, in Fig. 6.8, diverts current away from the base of Q_4 during all the time of turn-on for Q_4. This means that, to get the same amount of charge into the base as in the previous case, and in fact more charge is needed because this circuit allows Q_4 to saturate, we shall have to reduce the value of the final current in R_3 and make the final state P_1, shown in Fig. 6.12.

The advantage of the active pull-down is now obvious from Fig. 6.12. When Q_2 turns off, in both Figs. 6.8 and 6.11, Q_4 has a reverse base current,

Fig. 6.12.

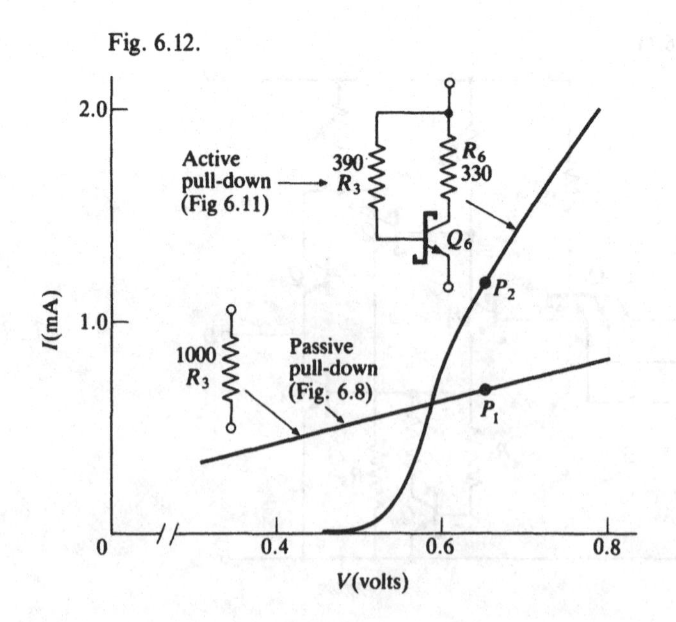

and in Fig. 6.11 this is nearly double the reverse base current which Q_4 has in Fig. 6.8. Coupled with the fact that Q_4 is a Schottky device in Fig. 6.11, this means that Fig. 6.11 is a very much faster circuit, if it is operated at the same current levels as Fig. 6.8, or that it will operate just as fast at very much reduced power level. We have here an interesting example of intentionally introducing a very non-linear element, the active pull-down, into a non-linear circuit, so that performance is greatly improved.

6.7 Emitter coupled logic (ECL)

The circuit shapes found in DTL and TTL appear to owe their origin to the idea of thinking in terms of simple switches. If this is so, it seems that there was a separate school of circuit designers who thought about computer circuits in a far more linear way: their ideas seem to come from the kind of circuits discussed in Chapter 3.

Fig. 6.13 shows an example of one of these circuits. Like Fig. 6.3, Fig. 6.13 is an IBM circuit, this time for the System/370 series of computers [15]. In Fig. 6.13, we recognise the long tailed pair discussed at length in Chapter 3, Fig. 3.2, as being made up of Q_3 on one side and several transistors, Q_1 and Q_2 only in this example, on the other side. The circuit has two states, depending on the input levels, in which the tail current [16] is switched either into Q_3 or into one or more of the input transistors.

Fig. 6.13.

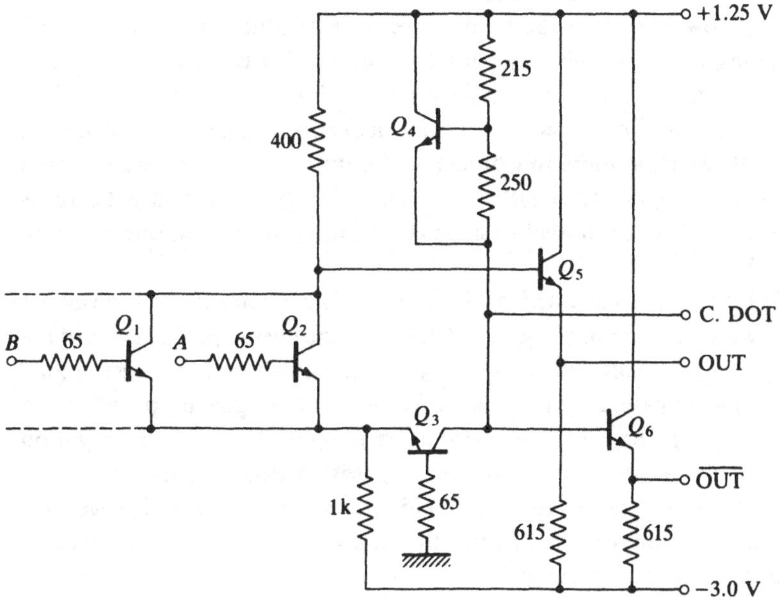

The resistor values are chosen so that the transistor collector voltages can never fall low enough to allow the transistors to saturate. This kind of logic circuit is, therefore, a very fast kind. Another great advantage of ECL is the fact that the current drawn from the power supply is nearly constant: a feature which underlines the *linear* origins of these circuit shapes.

An interesting feature of Fig. 6.13 is Q_4 and the two resistors associated with it. This is a voltage clamp, again a circuit shape of well established linear origin [17]. The current through the 250 Ω resistor must always be close to 0.65 V/250 Ω, which is 2.6 mA and very much greater than the base current of Q_4. Consequently this same 2.6 mA must flow in the 215 Ω resistor, setting the voltage across it and thus the voltage across all of this part of the circuit. This voltage clamp has been put in because the collector of Q_3 is brought out to a wiring point in this logic circuit for implementing *wired logic* functions. Wired logic simply means that other collectors, in other circuits, have their 'C.DOT' wiring points connected together and the voltage level at the C.DOT goes low if only one of the wired collectors draws current. A simple resistive collector load, like the 400 Ω which is shown in Fig. 6.13, would not allow this wired logic to be made and, at the same time, keep the required voltage level at the collectors, regardless of how many collector currents were on.

6.8 Very large scale integration (VLSI)

All the digital circuits which have been considered in this chapter up to now belong to the kind of technology which is concerned with building up large digital systems by connecting a very large number of separate devices together, each device containing just a few logic gates. This is why all the circuits had a particular final part which could be called an output stage: the interesting 'totem pole' output circuit, discussed at length in Section 6.5, and the emitter followers of Fig. 6.13. This output stage is needed to drive the fairly low impedance wiring which joins one circuit to another.

It is this intercircuit wiring which, as circuits become faster and smaller and work at lower power levels, determines the overall performance of the entire system. Somewhere in the system there will be the longest signal propagation path, a simple length of cable, l, with a signal propagating at a velocity, v, perhaps 60% of the velocity of light. When this propagation delay, l/v, becomes equal to the propagation delay through the logic circuits we have chosen to use to build up this digital system, then we have reached the limit of size, as far as the number of circuits is concerned, for that particular technology.

The way to get around this problem is to put as much of the digital system as possible on one very large silicon integrated circuit. This changes the circuit shapes which may be useful as logic gates because it is no longer necessary to have an output circuit capable of driving conventional wiring. As an example of what can be done along these lines it is interesting to look at some work published by IBM on one of the modules used in their computers of the early eighties [18]. This used silicon integrated circuits, each having 704 Schottky TTL gates, and each module contained 118 of these circuits mounted on a 15 cm × 15 cm ceramic carrier containing sixteen wiring planes to connect up the 118 integrated circuits [19]. The module, thus containing over 83 000 logic gates, dissipated a total of 300 W, which is 3.6 mW per gate. Gate propagation delays were just over 1 ns. This should be compared with an older circuit of about the same speed: Fig. 6.13. Fig. 6.13 involves a dissipation of 11 mW in the input part of the circuit and a further 40 mW in the two output emitter followers.

VLSI is also important as a technique for manufacture when the ultimate in speed and size is *not* the objective. Quite slow and rather simple digital processors can be made [20] as single integrated circuits and these will be not only cheaper but also far more reliable than systems which are built up from several smaller integrated circuits mounted upon one or more printed circuit boards.

Our interest in this book, however, is at the circuit level. How does the fact that one circuit can be connected to another circuit, perhaps only a fraction of a millimetre away, influence the circuit shape? What other changes do we find in the way digital circuit designers go about their work when they are concerned so intimately with the integrated circuit fabrication process itself?

One way of beginning to answer these questions is to go back to the start of this chapter and to Fig. 6.1. This shows the possibility of realising logic by means of a series combination of simple switches. To realise this kind of circuit on a microscopic scale the designer might think of the simplest of all arrangements: single transistors. Just the way that Fig. 6.1 is drawn makes us think of arranging a *series* combination of switches on a *surface*, and it is then a short step to think of *surface devices*: metal oxide semiconductor transistors, MOS devices.

In particular, the n-channel enhancement mode MOS transistor, and circuits made up from these devices alone are usually referred to as n-MOS, can be made with a very simple process on a p-type silicon substrate. Source and drain are diffused in as n-type, and then polycrystalline silicon is grown, by pyrolysis, on the very thin oxide which is over the channel area separating source and drain. This polycrystalline silicon is quite a good enough conductor to be used as the gate electrode and has the enormous advantage, over a metal, that silicon oxide is easily grown on top of it, so that connections to the gate can be made in polysilicon and these connections then oxidised and *crossed over* by the final metallisation pattern which connects sources, drains and, where necessary, polysilicon to metal. One of the best known texts on VLSI deals exclusively with these n-MOS circuits and systems [21].

Radically new circuit shapes are not found in digital n-MOS circuits. The transistors are used very much as switches and the resulting voltage level shifts and power losses are put right by means of simple amplifiers: very much like the way this was done in the bipolar circuits discussed here in connection with Fig. 6.3. A new feature in n-MOS, which is interesting, is the use of n-MOS transistors for the drain loads of other n-MOS transistors when these are used as amplifiers. This is done by using the load device in its ohmic region with gate and drain connected together. Analog circuits can also be built using this idea, so that n-MOS gives us the possibility of putting digital and analog circuits together on the same chip [22].

An advance upon n-MOS is CMOS: complementary MOS technology [23]. In this, we find a return to the kind of circuit shape shown in Fig. 6.5: two switches connected in series between a positive supply and ground. Fig. 6.14 shows how this can be realised using Q_1, a p-channel enhancement mode MOST, and Q_2, an n-channel enhancement mode MOST. This is a very interesting and new circuit shape, quite distinct from the bipolar circuit shape shown in Fig. 6.6 which, at first glance, seems to be very similar. Fig. 6.14 is also found as an output stage in operational amplifiers which use MOS devices of both polarities [24]. This circuit will be considered in more detail in Chapter 9.

Fig. 6.14.

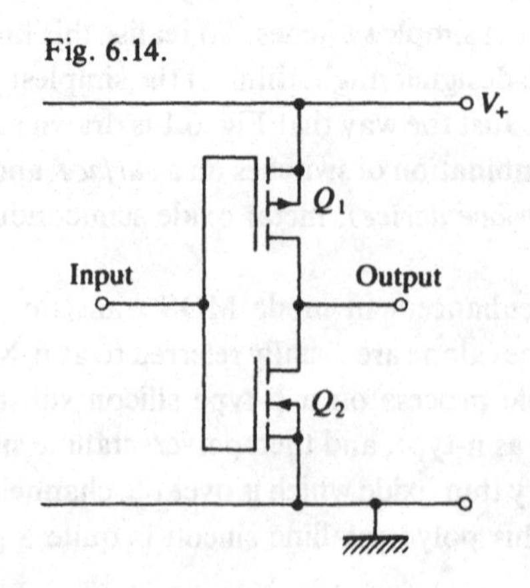

As part of a logic circuit, Fig. 6.14 has the advantage over n-MOS that current is only drawn from the power supply when the circuit is changing state. CMOS logic thus has a power dissipation which depends almost entirely upon the speed at which it is used. Again, polycrystalline silicon can be used to form the gates of the devices, and this allows effectively two levels of interconnection for these circuits.

All MOS VLSI circuits are remarkably simple and are also manufactured by a very simple process. This is a big factor in their commercial success because a simple process should give a high yield. The speed at which MOS VLSI circuits operate, particularly CMOS circuits, has increased considerably since they were introduced in the early seventies. This has happened because MOS speed depends far more than bipolar upon the reduction in scale which can be achieved as photolithographic technique improves. Propagation delays of only 1 ns have now been achieved with CMOS [25], showing that this technology has entered the speed range previously occupied by only bipolar silicon and field effect gallium arsenide circuits.

Bipolar VLSI remains the more interesting to the circuit designer, however, because it seems to have more potential for implementing a remarkable development in integrated circuit design: *merging*. This is explained in the next section, but note that merging, in a restricted sense, is also to be found in MOS technology [26].

6.9 Integrated injection logic

The circuits discussed in the previous section were introduced by asking the reader to glance at Fig. 6.1. Let us now look at Fig. 6.2 in contrast and imagine what this *parallel* combination of switches suggests on a microscopic scale. The answer could be that, as Fig. 6.1 suggested switches laid out on a surface, and thus surface devices, Fig. 6.2 suggests switches passing through a surface, and thus bulk devices. Unlike series switches, where access to both ends of the switch seems essential, parallel switches seem to imply that every switch has one end connected to a common ground and that the *circuit* consists of connecting up a pattern of switches on the top surface alone.

To realise Fig. 6.2, using solid state VLSI, we appear to need a switch with its control input brought to the surface and just one side of the switch also brought to the surface. How can this be done? The answer which immediately suggests itself is the bipolar transistor which, as an unisolated element in a silicon integrated circuit, has the appearance in cross-section shown in Fig. 6.15.

Fig. 6.15 shows how easy it would be to produce large quantities of bipolar transistors, by the conventional process, which had base and emitter connections freely available on the top surface of the silicon, while all the collectors of these transistors were connected via the n^+ substrate at the bottom of Fig. 6.15. Could useful logic be made by then wiring only on the top surface? Note at this point how we are beginning to think of new circuit shapes within the constraints of the process itself.

The answer to the question above appears to be negative. The first obvious problem is that it is the *collectors* of all these proposed transistors which have become common: we are more concerned with the collector as our output point. The second problem is that something is missing in the shape of a power supply. What kind of power supply should we use for logic which is going to be implemented by a *parallel* combination of switches? Fig. 6.2 shows the answer: a constant current generator.

The primitive circuit elements we need are thus shown in Fig. 6.16: bipolar transistors, with their *emitters* all in common, and constant current sources, all with one terminal taken to the same common bottom surface. Our VLSI system could then be realised by an interconnection pattern of metal on the top surface. In Fig. 6.16, the silicon surface is represented by the top horizontal line. The lower heavy horizontal line represents some

Fig. 6.15.

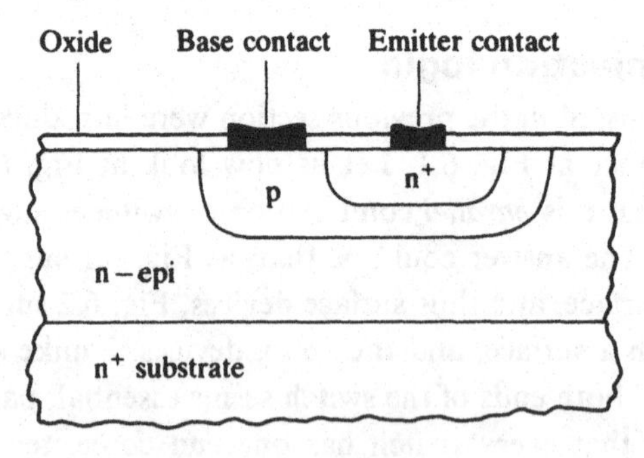

Fig. 6.16.

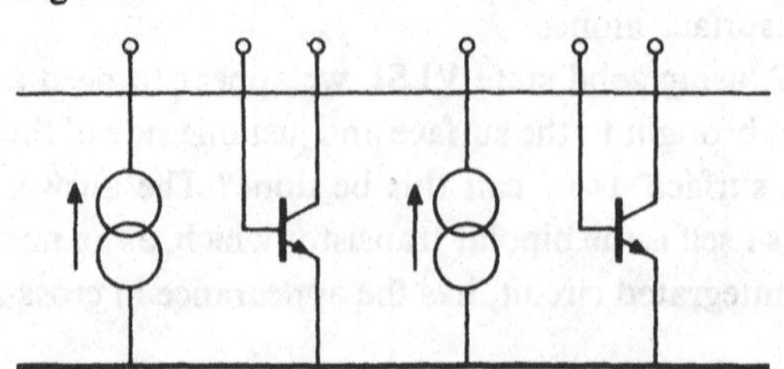

common layer below, perhaps an epilayer or the substrate. The transistors and generators are shown symbolically because we have not yet found out how to make these.

Note that in particular the direction of the current in the generators in Fig. 6.16 is upwards. This is a violation of a well established rule which the majority of circuit designers follow: positive rails at the top of the circuit, ground in the centre and negative rails at the bottom. However, this violation is helpful here because we are trying to find a circuit shape which is influenced by the fact that the circuit must be fabricated within the space near the top surface of a silicon wafer. We shall try to get the power supply lines underneath the top surface somehow.

How could the constant current generators, shown in Fig. 6.16, be realised in a bipolar process in a compact way? The answer can only be: as pnp transistors, the collectors of which are the output terminals of the constant current generators. This is shown in Fig. 6.17. The bases of all these pnp transistors can be taken to the same common ground point as the emitters of the npn devices. The emitters of the pnp transistors can also be connected to a common point and supplied with a single current source.

It is at this point that we can see how to make this circuit as a truly integrated circuit. A collector of a pnp transistor, shown in Fig. 6.17, will always be connected, via the metallisation pattern on the top surface of the wafer, to one or more npn transistor bases. There is no reason why this connection should not disappear, because the p-type silicon region acting as a collector for a pnp device can, in a different part of that p-type region, act as a base for an npn transistor. This is the idea of *merging*, mentioned at the end of the last section. Let us couple this idea with the fact that all the pnp transistors shown in Fig. 6.17 can be realised by means of one very large area pnp transistor having several isolated collectors. Such a transistor is shown in Fig. 6.18.

In Fig. 6.18 the p^+ substrate is the emitter, the n-type epilayer is the base and a second epilayer, of p-type, is the collector. This collector is divided up

Fig. 6.17.

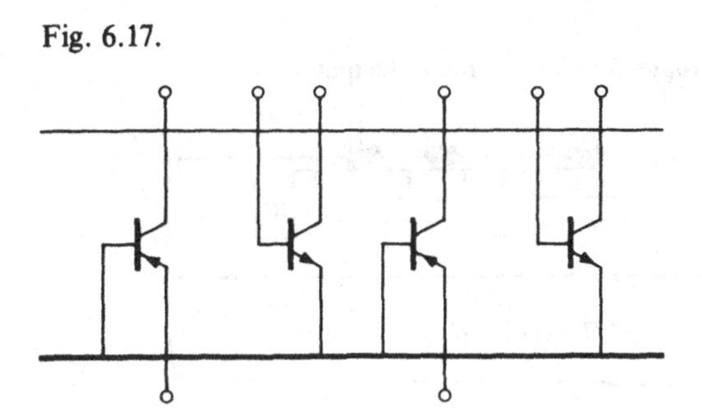

by a matrix of deep n^+ diffusion which passes from the top surface of the device, right through the p-type epilayer, into the n-type epilayer. Connection to this n^+ diffusion layer makes connection to the base of this large device, while connections to the, now isolated, p-type regions make connections to the several collectors.

Fig. 6.18 is drawn in cross-section. Looking down upon the top of the wafer we would see a rectangular grid of n^+ tracks. It is within this grid that we may now fabricate the npn transistors shown in Fig. 6.17. This is done by making an n-type diffusion into the p-type epilayer so that Fig. 6.18 is transformed into Fig. 6.19. Figs. 6.17 and 6.19 are really representations of the same thing: Fig. 6.17 is a circuit, Fig. 6.19 is a device geometry.

Fig. 6.19 shows a particular form of integrated injection logic, I^2L, which was first described by Blatt *et al.* in 1975 [27]. This was not the first kind of I^2L [28] but it is certainly one of the most interesting, from the point of view of merging: that is, making different parts of a circuit, which are connected together, by using different regions, of a particular diffusion, epilayer or substrate, of a device. The process required to fabricate the kind of logic shown in Fig. 6.19 is more complicated than that required for conventional bipolar digital circuits, and certainly more complicated than MOS or CMOS. For this reason, I^2L is found in rather special applications where its paeticular advantages justify extra cost [29]. There are a number

Fig. 6.18.

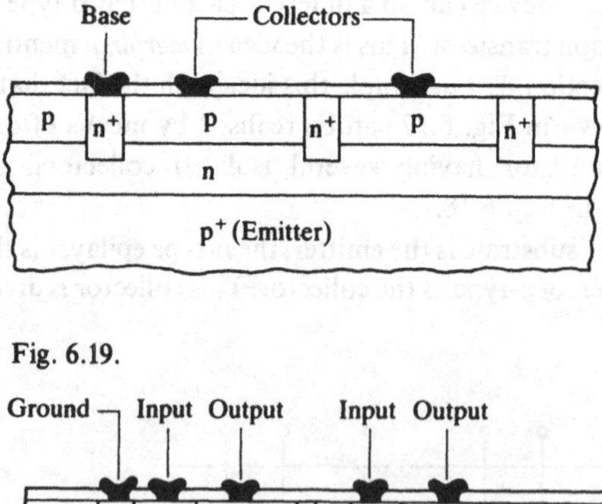

Fig. 6.19.

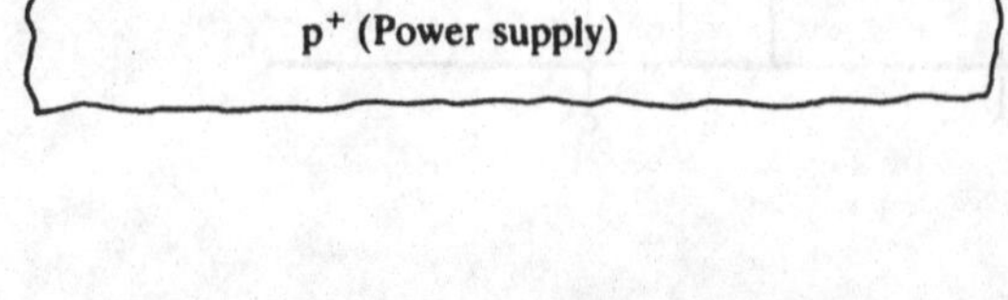

of very interesting versions of I^2L [30]. For example, the kind shown in Fig. 6.19 allows a large number of npn collectors to be formed inside one of the isolated regions of the p-type epilayer, and a single additional process step makes it possible to form several inputs to the bases of the npn devices via Schottky barrier diodes. These form multiple AND inputs. The circuit diagram of such a gate, having three such inputs and three collector outputs, is shown in Fig. 6.20. This has considerable system advantages [31].

I^2L is, of course, a saturating bipolar logic so that it will never be as fast as a non-saturating bipolar logic which is processed on the same scale. A good example of using I^2L to advantage has been published by Nakazato *et al.* [32], who describe a frequency divider for a 6 GHz input frequency. The first stages are ECL and then I^2L takes over at 580 MHz for the last stages, which output at 93.75 MHz.

6.10 An experimental circuit

This chapter has really been concerned almost entirely with integrated circuits, and it is difficult to propose any relevant experimental work without access to a process on which to try out some ideas. However, interesting experimental work can be done on digital circuits using discrete components, provided the speed of these circuits is kept low. It so happens that I^2L is ideal for this because it can be made to work over a very wide range of speed, depending upon the current levels which are set in the transistors.

Fig. 6.21 shows a very simple test circuit involving five I^2L gates of the simplest kind, simple inverters, connected as a ring oscillator. This is made up from ten discrete transistors, five npn and five pnp, and five discrete

Fig. 6.20.

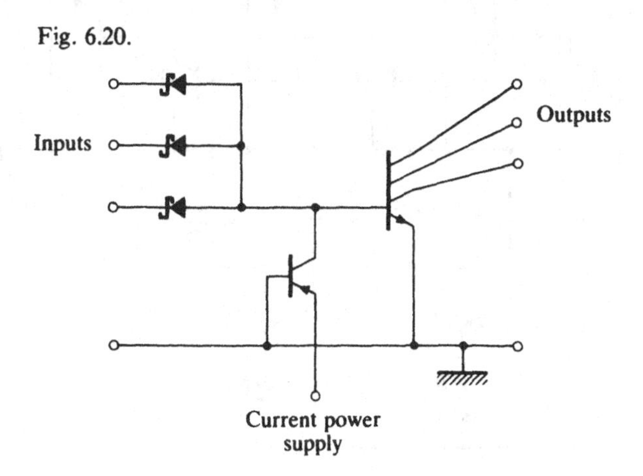

resistors. Transistors Q_1, Q_3, Q_5, Q_7 and Q_9 act as constant current sources, while Q_2, Q_4, Q_6, Q_8 and Q_{10} are the switches. The circuit has been drawn to suggest the way it should be laid out: the stray capacitance associated with each of the five gates should be about the same. This is discussed in more detail in the Appendix.

Testing logic gates by making ring oscillators is an excellent technique because each gate is driven and loaded by another gate of the same type. The circuit is being tested in a system environment. A single gate can be made to oscillate on its own, but this is only useful for gate circuits which are designed to drive coaxial cable [33] because a length of cable must be used which has a delay time well in excess of the gate propagation time. This is necessary because we need the oscillator to produce a square wave so that the rise and fall transients can really decay to the steady state after each transition. For any circuit that has VLSI potential, like I^2L has, a ring oscillator made up by a closed loop containing an odd number of gates is the only possibility because the capacitance of the interconnection, and thus its length, must be reduced to an absolute minimum. To obtain a good square wave, needed for the reasons given above, at least five gates should be used and seven is better [34].

Further discussion of Fig. 6.21 is given in the Appendix, along with some notes about the waveforms which can be observed.

Fig. 6.21.

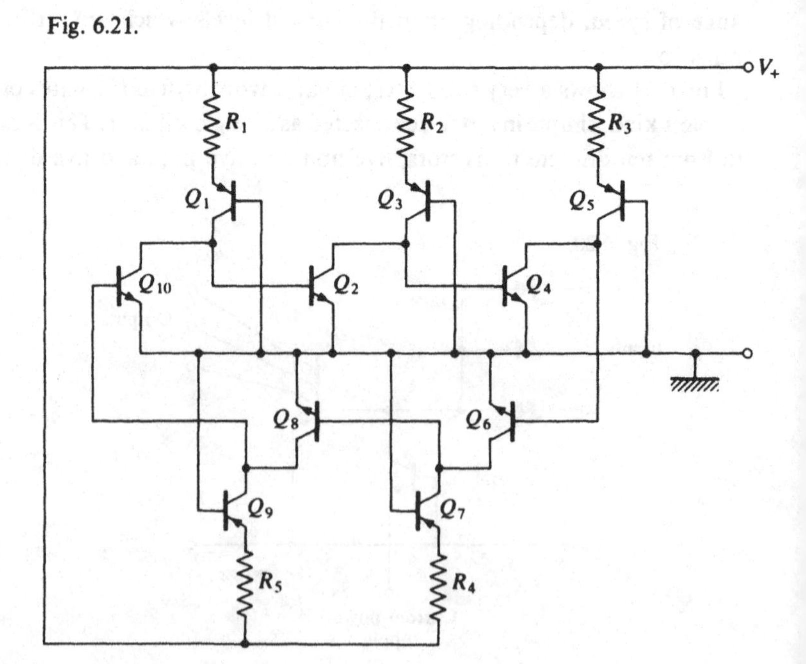

6.11 Conclusions

Rapid technical change in circuit design is nowhere more obvious than it is in the field of digital circuits. It is, therefore, all the more important to try to understand the reasoning behind the continuous changes and developments which are recorded in the literature. Some changes come from a better understanding of circuits and circuit design or a better understanding of new devices. Some changes seem to come more from developments in the processes which become available. These two aspects are intimately connected but we may have an illustration of them both by looking at two circuits from this chapter.

If the circuit shown in Fig. 6.3 is compared with the circuit shown in Fig. 6.20, two features come across to us as a result of the years of development and thought which separate these two logic gates. Functionally, at least if Fig. 6.20 had only one output, the two circuits are not all that different.

The first feature is that the older circuit has been designed with the idea of setting voltage levels at inputs and outputs very much as a first consideration. It is the voltage level which represents the logic state, '1' or '0', and it is the voltage level which actually causes the diode switches to be on or off. The more recent circuit, Fig. 6.20, is much more subtle in that it does have a one to one correspondence to the primitive kind of logic shown in Fig. 6.2. Voltage levels are not so much the designer's concern. It is the *state* of the circuit which is important, and, of course, the state of the circuits which are connected to its inputs.

The second feature which stands out, when we compare Figs. 6.3 and 6.20, is more down to earth and is a question of complexity and scale. Fig. 6.3 involves nine discrete components: five diodes, three resistors and a transistor. Even if Fig. 6.3 was made as an integrated circuit, and there is no reason why it should not be, it would have to have at least six isolated regions: one for diodes D_1 to D_4, one for diode D_5, three for each of the three resistors and, finally, a sixth region for the transistor, Q_1. Contrast this with Fig. 6.20. As Fig. 6.19 shows, all of Fig. 6.20 can be realised inside one of the areas of the p-type epilayer which is surrounded by a grounded n^+ diffusion. There would be three Schottky barrier inputs and three n-type diffused areas for the three collector outputs. This could be an area much less than 100 μm square. The rest of the circuit is realised by the structure which lies underneath the top surface: the vertical structure which we see in Fig. 6.19, built up through the thickness of the device. This is a revolution in technique when it is compared to the simple integrated circuits discussed in the earlier sections of this chapter. Integrated injection logic is an example of a really integrated form of solid state circuitry.

Notes

1 B. Randell, *The Origins of Digital Computers*, Springer Verlag, 1973, pp. 287–92.
2 W. Renwick, *J. Brit. IRE*, **20**, 563–72, 1960.
3 K.-H. Czauderna, *Konrad Zuse, der Weg zu seinem Computer Z3*, R. Oldenbourg Verlag, 1979.
4 E.M. Davis, W.E. Harding, R.S. Schwartz and J.J. Conway, *IBM J. Res. Dev.*, **8**, 102–14, 1964. This paper gives many photographs of the circuits showing how these were made using 'flip-chip' devices. This technique of automated production developed tremendously at IBM: from devices mounted on just three ball contacts in 1964, twenty contacts were being used in 1971 (*IBM J. Res. Dev.*, **15**, 384–90, 1971) to mount integrated circuits containing about eight gates each. This was for System/370. By 1982, IBM were using this technique with chips having 704 gates mounted on 132 ball contacts (*IBM J. Res. Dev.*, **26**, 2–11, 1982). The best photographs of these remarkable VLSI circuits, for System/38, are to be found in *Electronics*, **52**, No. 6, 107, March 15 1979. This continuity in manufacturing technique may be a vital factor in understanding the outstanding success of a particular organisation. As Freeman has written in his introduction to G. Dosi's important work, *Technical Change and Industrial Transformation: the theory and an application to the semiconductor industry*, Macmillan, 1984, 'it is both empirically ridiculous and theoretically untenable to start from the assumption that all agents are equal in their access to technology [manufacturing technique: which depends upon *people*, how they are cared for and trained so that they can design and build the machines needed for highly automated production] in any branch of industry and in their capacity to innovate'. The internal state of any organisation from this point of view is very difficult to determine and Dosi's work represents one of the most interesting beginnings in this new field of research. As far as IBM is concerned, a very interesting paper by W.E. Harding, 'Semiconductor manufacturing in IBM, 1957 to the present: a perspective', *IBM J. Res. Dev.*, **25**, 647–58, 1981, gives quite detailed information for the earlier years and clearly confirms the fact that IBM was, by far, the biggest manufacturer of semiconductor devices and also the only one which used complete automation of all stages of manufacture and testing, locating its plants in the most developed countries: USA, France and Germany. The book *IBM and the US Data Processing Industry*, by F.M. Fisher, J.W. McKie and R.B. Mancke, Praeger, New York, 1983, pp. 339–41, is a further valuable source of information and confirms the importance of continuity in technique, mentioned above.
5 A very early reference to this is P.M. Thompson, *Electronics*, **36**, No. 37, 25–9, September 13 1963.
6 Chapter 3, note 9.
7 An interesting historical paper, which gives the circuits of Fig. 6.4 and Fig. 6.7, is by D. Christiansen, *Electronics*, **40**, No. 5, 149–57, March 6 1967. This paper is remarkable for the excellent colour photographs of the devices of that time.

8 No detailed treatment of the dynamics of Fig. 6.8 was published until 1968: P.J. Tizzard and M.J. Turner, *Microelectronics*, **1**, No. 2, 28–33, January 1968.
9 The data for Fig. 6.9 come from the US Market Reports and Forecasts which are published in January every year by *Electronics*. The raw data have been used because the price per circuit fell over the period: correcting for this would give a plot of the number of devices sold and would tilt the whole curve upwards. On the other hand, 'inflation' was quite marked over the period: correcting for this would reduce the asymmetry which Fig. 6.9 shows.
10 Note 8 above.
11 W. Schottky published his fundamental work on metal–semiconductor junctions in 1939. He worked at Siemens in Germany. See H.C. Torrey and C.A. Whitmer, *Crystal Rectifiers*, McGraw Hill, 1948, p. 78.
12 M.I. Elmasry, *Digital Bipolar Integrated Circuits*, Wiley Interscience, 1983, Figs. 2.7 and 3.1.
13 R.H. Baker, *Electronics*, **30**, No. 3, 190–3, March 1 1957.
14 Note 12 above, Fig. 4.12.
15 P.E. Fox and W.J. Nestorix, *IBM J. Res. Dev.*, **15**, 384–90, 1971. See note 4 above.
16 The logic levels at the output of the circuit shown in Fig. 6.13 are about +0.6 V and −0.6 V. Consequently the current in the 1000 Ω tail resistor is either about 3 mA or 2.3 mA.
17 The basis of the voltage clamp is the idea of defining a current in a resistor by connecting the emitter junction of an active transistor across it. This was discussed in Section 3.3.
18 M.S. Pittler, D.M. Powers and D.L. Schnabel, *IBM J. Res. Dev.*, **26**, 2–11, 1982. See note 4 above.
19 A.J. Blodgett and D.R. Barbour, *IBM J. Res. Dev.*, **26**, 30–6, 1982.
20 The most familiar examples of these digital VLSI devices are the calculator, microprocessor and microcontroller chips which began to be used in the early seventies. An excellent review of this development is the article 'Microprocessors in brief', by R.C. Stanley, *IBM J. Res. Dev.*, **29**, 110–31, March 1985.
21 C. Mead and L. Conway, *Introduction to VLSI Systems*, Addison-Wesley, 1980.
22 Chapter 12 of the second edition of *Analysis and Design of Analog Integrated Circuits*, by P.R. Gray and R.G. Meyer, John Wiley, 1984, is called 'MOS Amplifier Design', and deals with analog circuits using CMOS and n-MOS. There is also mention of switched capacitor technique, which is particularly important in this kind of MOS technology where capacitors having values with very precisely determined ratios may be fabricated.
23 D.L. Wollesen, *Electronics*, **52**, No. 19, 116–23, September 13 1979.
24 O.H. Schade, *RCA Rev.*, **37**, 404–24, Figs. 4 and 5, 1976.
25 Y. Takayama, S. Fujii, T. Tanabe, K. Kawauchi and T. Yoshida, *IEEE ISSCC 1985 Digest of Technical Papers*, pp. 196–7.
26 M.I. Elmasry, *Digital MOS Integrated Circuits*, IEEE Press, 1981, p. 73.

27 V. Blatt, P.S. Walsh and L.W. Kennedy, *IEEE J. Solid State Circuits*, **SC–10**, 336–42, 1975.

28 I^2L appears to have been invented in 1972 by H.H. Berger and S.K. Wiedmann (*IEEE J. Solid State Circuits*, **SC-7**, 340–6, 1972), and by K. Hart and A. Slob (same issue of this journal, pp. 346–51). A very interesting issue of *Electronics* (**48**, No. 21, 66–9, October 16, 1975) gives the biographies of the four inventors and some history of the process of invention.

29 *IEEE Spectrum*, **22**, No. 1, 26, January 1985.

30 See Chapter 5 of note 12 above.

31 Fig. 4 of the paper referred to above as note 27 makes this particularly clear.

32 K. Nakazato, T. Nakamura, J. Nakagawa, T. Okabe and M. Nagata, *IEEE ISSCC 1985 Digest of Technical Papers*, pp. 214–15.

33 R.F. Sechler, A.R. Strube and J.R. Turnbull, *IBM J. Res. Dev.*, **11**, 74–85, 1967.

34 R. Müller and J. Graul, *IEE Conf. Pub.* No. 130, 30–1, September 1975.

7

Sinusoidal oscillators

7.1 Origins

In this chapter we shall look at circuits which are designed to be sinusoidal signal sources. Relaxation oscillators, which produce non-sinusoidal waveforms, have already been considered, briefly, in Chapter 1, Figs. 1.1, 1.2 and 1.3, and also in Chapter 6, Fig. 6.21.

The simplest electronic oscillators involve a single active amplifying device and some kind of resonator. In general, the circuit shape always takes the form shown in Fig. 7.1. The active device, in this example a bipolar transistor, has an impedance, Z_1, connected across its input, an impedance, Z_2, connected across its output, and an impedance, Z_{12}, which couples these input and output circuits together. An example of such a simple oscillator has already been considered in Chapter 2. Fig. 2.4 was found to be an oscillator under certain conditions. In that circuit, Z_1 and

Fig. 7.1.

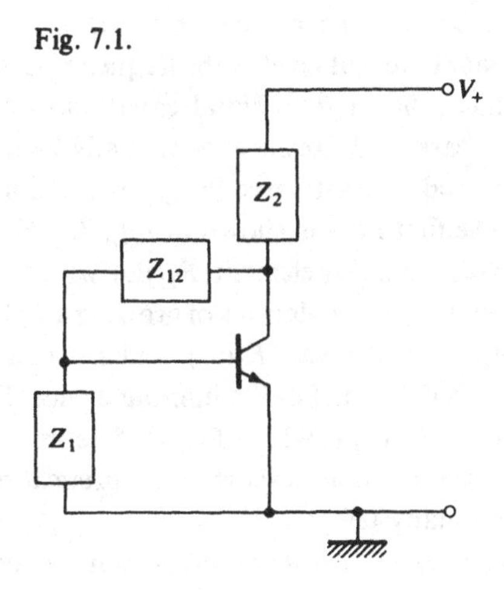

Z_2 were the parallel tuned circuits, L_1C_1 and L_2C_2, while Z_{12} was the feedback capacitance, C_{rs}. In the amplifier application, discussed in Chapter 2, C_{rs} was minimised as much as possible. To make an oscillator, the feedback is deliberately arranged so that the circuit oscillates and the amplitude of oscillation is then either determined by the supply voltage, V_+ in Fig. 7.1, or limited by the maximum current that can flow in the active device.

Simple oscillator circuits of this kind may, of course, use the active device in common base or common collector connection, the possibilities shown in Figs. 2.1(*b*) and 2.1(*c*), and the impedances, Z_1, Z_2 and Z_{12}, may be simply inductive or capacitive. The oscillator circuits are then often given names: Hartley [1], Colpitts, Clapp and so on, after the patentees, but the circuit shape is always the same as that shown in Fig. 7.1. An excellent paper by Baxandall [2] reviews the transition between these early circuits and the kind of oscillator circuit which is used today.

7.2 Oscillators as systems

Very simple oscillator circuits using a single active device, or perhaps a single simple integrated circuit, are still very widely used as local oscillators in radio and television receivers. More and more, however, we find digital techniques being used to synthesise [3] a signal of some desired frequency, and, of course, any desired waveform. These synthesisers call for one or more crystal controlled oscillators to provide their foundation sources. Crystal oscillators can involve some interesting circuit problems and these will be considered at the end of this chapter.

Conventional variable frequency oscillators are best looked upon as feedback systems. These must be non-linear because a useful oscillator should give a well defined, constant, output level as the frequency is varied. A linear system, one which has a linear differential equation, can only generate oscillations which increase or decrease exponentially with time.

From this point of view we find at least three fairly distinct kinds of system shape for oscillators. The first kind is shown in Fig. 7.2. Here we have a feedback loop which involves a linear element, $F_1(\mathrm{j}\omega)$, which, at low level, produces an oscillator loop to give oscillations of growing amplitude. The loop also contains a non-linear element, $F_2(v_1)$, which has a gain, $\mathrm{d}v_2/\mathrm{d}v_1$, that *falls* with increasing v_1: some kind of limiting device. $F_1(\mathrm{j}\omega)$ thus determines the oscillator frequency while $F_2(v_1)$ determines the amplitude. Sidorowicz [4] has published an important theoretical review of this type of oscillator, giving many references.

The second kind of circuit shape, or system shape, an oscillator may have is shown in Fig. 7.3. Here, there is a positive feedback path, via R_1 to $Z(\mathrm{j}\omega)$, which makes the system start oscillations of growing amplitude. There is also a negative feedback path, via R_3 to R_2, which involves a non-linear resistor, $R_3(i)$. This non-linear resistor has a value which *falls* with increasing current: it could be a thermistor.

A suitable choice of the four components shown in Fig. 7.3 will produce an oscillator in which the amplitude of oscillation will grow when the circuit is first switched on, and then settle down to some constant level. This kind of oscillator is often referred to as a Meacham bridge circuit [5] because R_1, R_2, $R_3(i)$ and $Z(j\omega)$ form a classical bridge circuit which is driven by v_{out} and has its out-of-balance voltage across the input of the amplifier, A. The circuit can thus be looked upon as a self-balancing bridge circuit, and it could be that this reasoning led to the circuit shape.

The third kind of oscillator we shall find in practice is more sophisticated and is shown as a block diagram, or system shape, in Fig. 7.4. Here we find the same oscillator loop as in Fig. 7.2 but the non-linear element, $F_2(v_3)$, now has its gain, dv_2/dv_1, controlled by a separate voltage input, v_3. This gain block, $F_2(v_3)$, could, for example, be simply a dual gate MOST, of the kind used in Fig. 2.4, connected as a common source amplifier and having v_1 taken to gate 1 and v_3 taken to gate 2. The control voltage, v_3, is produced by the amplifier, A_2, in Fig. 7.4, which compares the rectified

Fig. 7.2.

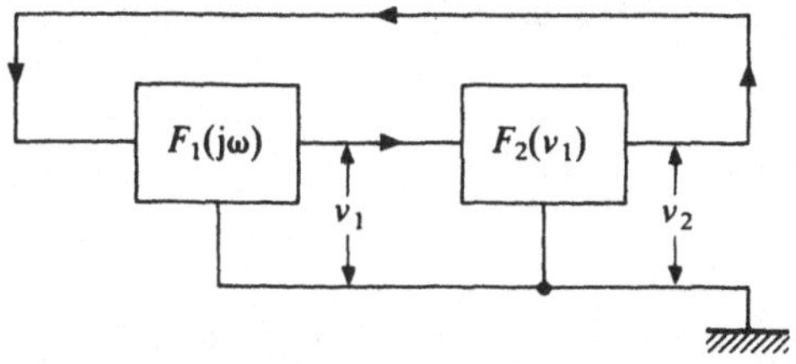

Fig. 7.3.

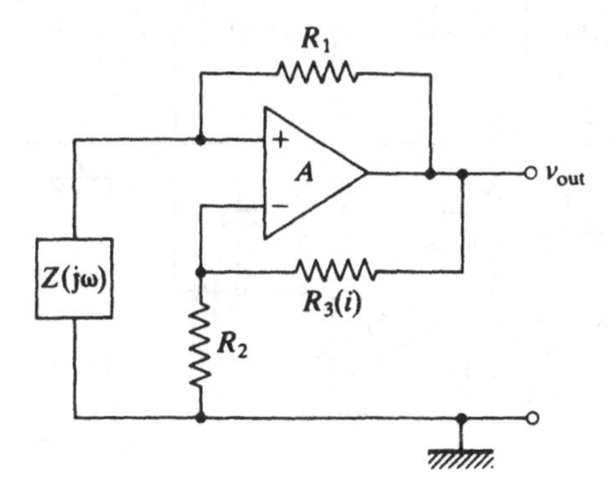

output of the oscillator with a manually set voltage level. The design of this kind of oscillator system is particularly interesting because it involves a non-linear control loop which must be stable and have the correct kind of transient response to changes that take place when the oscillator is tuned. A major factor in this stability problem is the rectifier time constant: C_1R_1 in Fig. 7.4.

7.3 Varying the oscillator frequency

An oscillator that is used for testing or for experimental work needs to be tunable over a wide range. This requirement may not affect the circuit shape but it will determine which components the designer chooses to make variable. The very simple oscillator circuit of Fig. 7.3 can be used to illustrate this.

Fig. 7.3 has been redrawn in Fig. 7.5 with the details of a suitable $Z(j\omega)$ included. This oscillator circuit is so simple that we may write down the complete differential equation for very small amplitude oscillations. This will tell us how the oscillator would start oscillating, and also tell us the conditions needed for this to happen.

For very low level oscillations we may assume that $R_3(i) = \infty$ and concentrate upon the positive feedback loop. Assuming also that A is large, so that R_1/R must be large, a current v_{out}/R_1 may be assumed to flow in R_1 and thus

$$v_{out}/R_1 = v/R + C(dv/dt) + i_L \tag{7.1}$$

Fig. 7.4.

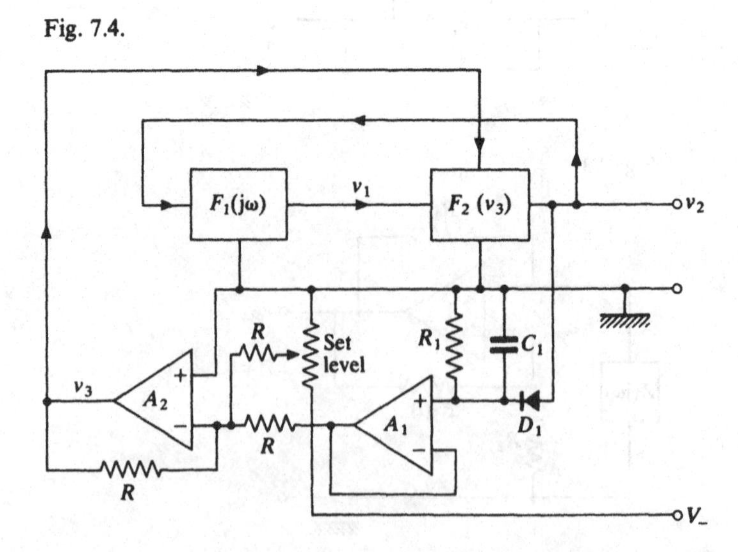

where v is the voltage across the amplifier input and i_L is the current flowing in the inductor L. Substituting $v_{out} = Av$ into equation (7.1), and differentiating, gives

$$d^2v/dt^2 + (1/CR - A/CR_1)dv/dt + v/LC = 0 \qquad (7.2)$$

Equation (7.2) is a simple second order differential equation for v and has a solution giving oscillations at a frequency $\omega = 1/\sqrt{LC}$. These grow exponentially provided the coefficient of dv/dt is negative. This tells us that we must ensure that $(A/R_1) > (1/R)$ if the oscillator is to start oscillating. It also tells us that this condition does not depend upon the oscillation frequency: $\omega = 1/\sqrt{LC}$ may be varied by varying L or C and the condition for oscillation does not appear to be affected. However, this conclusion is not really valid because R represents the losses in L and C as well as any resistance which might be deliberately included in the design. These losses will vary with frequency.

Another point which may be made from the very simple oscillator circuit of Fig. 7.5 concerns the arrangement of the components which would be

Fig. 7.5.

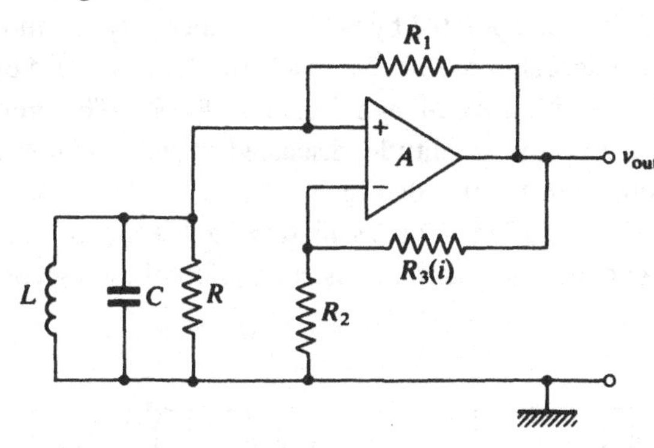

Fig. 7.6.

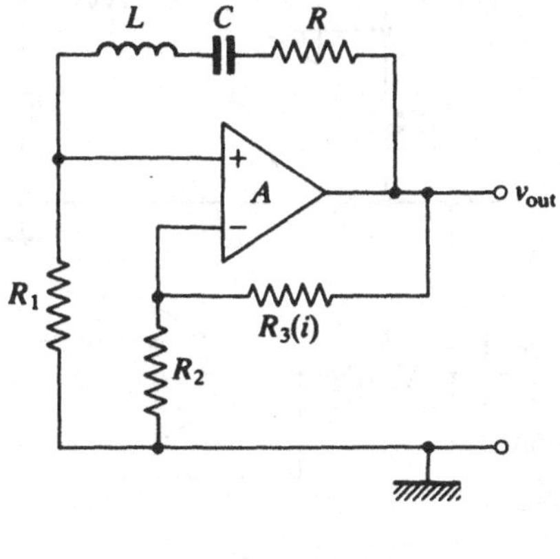

made variable. In Fig. 7.5 these are either L or C, and one side of these components is taken to ground. An alternative oscillator circuit is shown in Fig. 7.6. This has exactly the same circuit shape as Fig. 7.5, and the differential equation may be written down just as easily, but Fig. 7.6 is not a good circuit for a variable frequency oscillator because both L and C are floating. The mechanical construction of any variable component implies that one side of it will have a large capacitance to ground. This is the side that should be grounded.

7.4 Oscillators using resistance capacity (RC) networks

The restrictions and influences upon the choice of circuit shape, discussed in the previous section for very simple oscillator circuits using *LCR* resonators, become more subtle when oscillators which use only *RC* networks are considered. *RC* oscillators are more suitable for low frequencies and have advantages when very wide tuning range is needed.

One of the most familiar *RC* oscillators is shown in Fig. 7.7. This has three phase-lag *RC* sections, separated by buffer amplifiers, A_1, A_2 and A_3, which produce a total phase shift of 180°. The oscillator loop is closed by an inverting amplifier, A_4, with a small signal gain of R_2/R_1. The limiting action needed for defining the amplitude, discussed above in connection with Fig. 7.2, is assumed to be included in A_4.

When the system shown in Fig. 7.7 is oscillating, each *RC* section must introduce a phase-lag of 60° and, as the transfer function of each section is

Fig. 7.7.

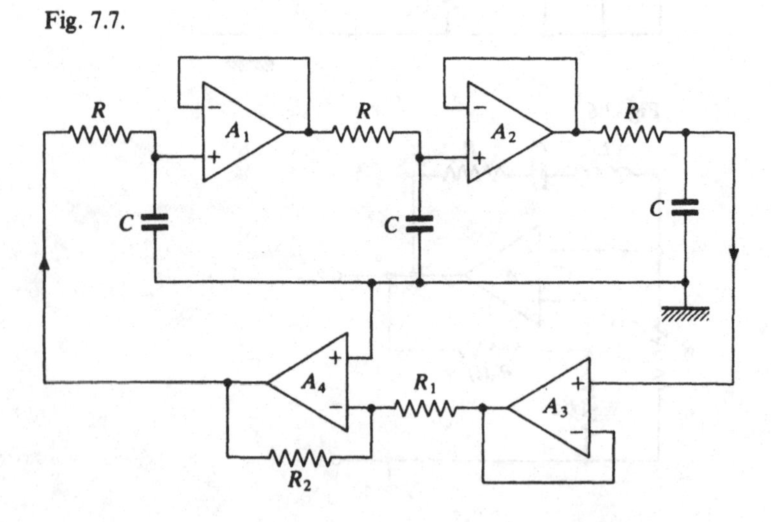

$(1+j\omega CR)^{-1}$, this means that the frequency of oscillation must be

$$\omega_0 = \sqrt{3}/CR \tag{7.3}$$

It also follows that the gain of each *RC* section is 0.5, so that the gain of A_4, R_2/R_1, must be greater than 8.0 if the circuit is to oscillate.

Now the interesting point here is what is needed to vary the frequency of this oscillator. As all the capacitors in Fig. 7.7 have one side grounded, it would seem sensible to vary these: equation (7.3) shows that ω_0 varies as $(C)^{-1}$. We must remember, however, that *all three* capacitors must be varied together. If only one is varied, the change in frequency is quite small, ω_0 varies as $(C)^{-1/3}$, and, far more important, the gain needed in A_4 to sustain oscillations *depends upon frequency*. This is why this kind of oscillator circuit is rarely found in practice.

7.5 All-pass networks in oscillators

A good *RC* oscillator needs a network which gives a larger phase shift than can be obtained with a single RC section. Another useful feature would be constant gain as the frequency is changed. A network which has this property is usually called 'all-pass'.

The importance of all-pass networks for *RC* oscillators has been known for a long time [6] and these networks are the origin of some very interesting circuit shapes. The fundamental idea of an all-pass network is shown in Fig. 7.8. Two inputs are shown, one of which is simply the inverse of the other, and the output voltage is then given by

$$v_{out} = v_{in}(1-j\omega CR)/(1+j\omega CR) \tag{7.4}$$

Equation (7.4) shows that, as the frequency is varied from zero to infinity, the phase angle between v_{in} and v_{out} varies from zero to 180°, but v_{out} stays equal in magnitude to v_{in}. To be precise, the relationship between

Fig. 7.8.

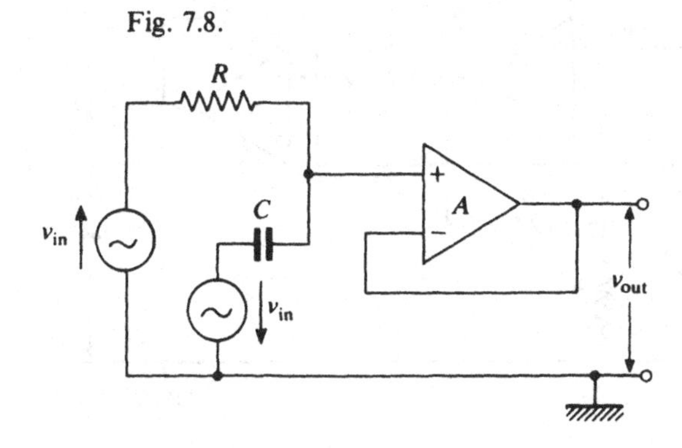

v_{out} and v_{in} is simply

$$v_{out} = v_{in} \exp(-2j\phi) \tag{7.5}$$

where

$$\phi = \arctan(\omega CR) \tag{7.6}$$

Two networks of the kind shown in Fig. 7.8 would be all that would be needed, together with an inverting amplifier, to make an *RC* oscillator running at a frequency $\omega_0 = 1/CR$. However, the form of the circuit shown in Fig. 7.8 is not very practical, particularly in view of the fact that neither *C* nor *R* have one end grounded.

Fig. 7.9 shows a way in which the all-pass transfer function of equation (7.4) may be realised so that one end of the resistor, *R*, is grounded [7]. The other components in this circuit are two identical resistors, both labelled R_1, and the two operational amplifiers, A_1 and A_2, with four resistors, also all identical, all labelled R_2. Connected in this way, A_1 and A_2 form a well known differential amplifier with a differential gain of -2. This means that $v_{out} = -2(v_1 - v_2)$, where v_1 and v_2 are shown in Fig. 7.9. As $v_1 = v_{in} j\omega CR/(1 + j\omega CR)$ and $v_2 = v_{in}/2$, it follows that v_{out} is given by equation (7.4).

It is possible to follow the evolution of the practical all-pass circuit, Fig. 7.9, from what might be called its academic prototype, Fig. 7.8. The need to ground one end of *R*, because *R* is to be the variable frequency control, means that the voltage at the junction of *R* and *C*, which was taken relative to ground in Fig. 7.8, must now be taken relative to some fraction of v_{in} and inverted. The fraction of 1/2 is the most easily arranged and the differential amplifier of Fig. 7.9 allows the voltage at the junction of *C* and *R* to be transferred relative to this new level.

Fig. 7.9.

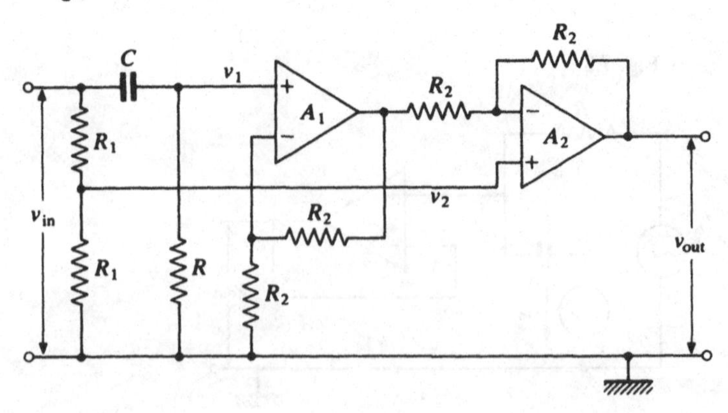

An oscillator which will have a very wide range of tuning can be made by connecting two of the circuits shown in Fig. 7.9 in cascade and closing the oscillator loop with an inverting amplifier having a small signal gain just above unity. Tuning is done by means of a two-gang potentiometer. It is interesting to note that the all-pass characteristics of the phase shift networks make close tracking of the variable components less critical than in an *RC* oscillator of the kind shown in Fig. 7.7. Similarly, only one component may be varied, to give a smaller frequency variation, proportional to $(R)^{-1/2}$ instead of $(R)^{-1}$, because the gain needed around the oscillator loop will remain the same at all settings of the variable components.

One thing which is more difficult, with oscillators using all-pass networks, is the reduction of harmonics in the output sine wave. These are introduced by the non-linear elements controlling the final steady state amplitude of the oscillator. The all-pass nature of the oscillator loop means that there is no built in filtering of these higher frequencies. This is why all-pass oscillators usually use a control loop for amplitude stabilisation, like the one shown in Fig. 7.4. A particularly interesting design has been published by Mayer [8], which has both amplitude and phase control loops. Mayer's design uses photon coupled field effect transistors as controlled resistors.

7.6 An experimental RC oscillator

As the first experimental circuit for this chapter, let us look at an *RC* oscillator design which is somewhat specialised: a fixed frequency oscillator intended for very low frequency work and giving a very pure sine wave output [9].

The circuit diagram is shown in Fig. 7.10 and more detail is given in the Appendix. This oscillator is one of the kind first shown in Fig. 7.2. The amplifiers A_1 and A_2, with their associated feedback components, form the element $F_1(j\omega)$, of Fig. 7.2, while the non-linear element $F_2(v_1)$ is formed by R_5 and a pair of Zener diodes connected back to back. Amplifier A_3 simply gives a buffered output.

The oscillator shown in Fig. 7.10 should give a very pure sinusoidal output because the limiting which is introduced by R_5, D_1 and D_2 distorts the signal arriving at the input end of R_1 symmetrically. This means that only odd harmonics of the fundamental should be present. From the input end of R_1 to the output of A_2, the oscillator loop is designed to have a very strong, -18 db/octave, low-pass characteristic. It follows that the third and higher harmonics, generated by R_5, D_1 and D_2, should be strongly attenuated.

To see how this is done, note that the first amplifier, A_1, is connected as a two-pole low-pass filter, and, if we make $R_1 = R_2 = R$ and $C_1 = 2C_2 = 2C$, the transfer function

$$v_2/v_1 = 1/[1+\sqrt{2}\mathrm{j}(\omega/\omega_0) - (\omega/\omega_0)^2] \tag{7.7}$$

where

$$\omega_0 = 1/\sqrt{2}CR \tag{7.8}$$

is maximally flat [10].

This choice of a maximally flat filter means that, at the frequency ω_0, where equation (7.7) shows that the filter introduces a phase lag of 90°, the attenuation introduced is $\sqrt{2}$. An oscillator loop can now be made by adding the inverting integrator, A_2 in Fig. 7.10, which has a phase lag of 270°, and choosing C_3R_3 so that the gain of the integrator,

$$v_{\mathrm{out}}/v_2 = -1/(\mathrm{j}\omega C_3R_3) \tag{7.9}$$

has a magnitude just a little greater than $\sqrt{2}$. This, of course, assumes that $R_5 \ll R_1$.

Fig. 7.10 is a very interesting circuit to experiment with because, if the operating frequency is made fairly low, for example about 1 Hz, the measured frequency and amplitudes may be compared with the simple theory given here. Then, the full third order differential equation may be derived and the time constant which dominates the build up of oscillations observed and checked. It is also interesting to make the oscillation frequency first as low, and then as high, as possible, and to understand what limits this range in this particular circuit. These measurements are discussed in the Appendix. This oscillator circuit is, of course, not one which lends itself to a simple variable frequency control: the loop gain has

Fig. 7.10.

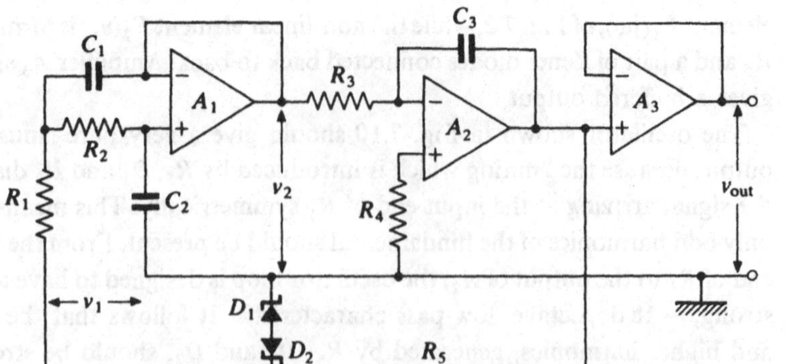

been made very frequency dependent, intentionally, so that a good sinusoidal waveform is obtained.

7.7 Crystal controlled oscillators

At radio frequencies, the best possible signal source, if accuracy and stability are the prime requirements, is a quartz crystal oscillator. Such oscillators have been used in electronics for a very long time [11]. The quartz resonator is usually represented in the literature of electronics by the equivalent circuit shown in Fig. 7.11. This conceals some very complex physical phenomena. For a typical quartz resonator near 1 MHz, the 'inductor', L_1, would be about 3 H, the 'capacitor', C_1, would be about 0.01 pF and the 'resistance', R_1, about 100 Ω. The capacitance C_2 is very close to the simple electrical capacitance of the crystal, which would be in the form of a thin slice, about 1 mm thick, with electrodes upon either side. C_2 is thus a few pF.

It follows that a quartz crystal resonator apparently has two resonant frequencies: a series resonance very close to $\omega = 1/\sqrt{L_1 C_1}$ and a parallel resonance just above $\omega = 1/\sqrt{L_1 C_1}$, which may be tuned very slightly by adding extra external capacitance to C_2. The physics of these two resonances is most interesting and, for clarity, Nye's text [12] has not been improved upon by more recent books. The point is that the mechanical stress in the quartz crystal is determined by the electric field, while the electric polarisation of the quartz is mainly determined by its state of strain. Under resonant conditions, stress and strain do not vary in phase: strain lags stress because of the inertial mass of the material. It follows that the well defined mechanical resonance is found when the voltage across the crystal is quite small, but the current through it, which is simply the rate of change in electric polarisation, is at a maximum. This is the series resonance of the equivalent circuit shown in Fig. 7.11, and series resonance is used in most oscillator circuits which aim for high stability.

Fig. 7.11.

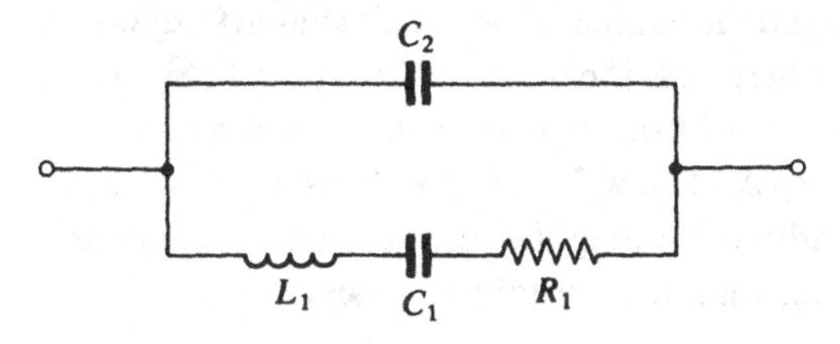

7.8 Crystal oscillator circuits

The design of crystal oscillator circuits has been discussed by Baxandall [13] in a paper which is essential reading for circuit designers. Baxandall was among the first people to point out 'the influence exerted by the manner of drawing the circuit diagram on the way one thinks about the functioning of the circuit' [14].

The most important point that the designer must keep in mind when considering a circuit for a crystal oscillator is the very large Q factor of the quartz resonator. This can be above 10^5, and this has two consequences: it is not difficult to drive the crystal into its non-linear elastic region, and, secondly, a small change in the phase characteristics of the oscillator circuit can still cause a large, by quartz oscillator standards, shift in oscillator frequency.

Consider the problem of overdriving the crystal first. The Q factor of any resonator is a measure of the energy stored in the resonator divided by the energy lost per cycle. For example, a 1 MHz crystal which is driven so that the dissipation within the crystal is 10 mW, a typical value for many published circuits [15], is losing $10 \times 10^{-3} \times 10^{-6} = 10^{-8}$ joules per cycle. If the Q factor of the crystal is 10^5, this means that we have a maximum of 10^{-3} joules stored in the elastically distorted quartz resonator. What does this really mean? Stored in the elastic deformation of a few cubic millimetres of quartz, 10^{-3} joules is rather an impressive amount of energy. It is nearly as much as the energy stored in a 1 μF capacitor charged up to 50 volts, and it is more than that in the impact of a weight of 1 gram falling from 10 cm. If more power than this 10 mW is dissipated, it is quite possible that the crystal will fracture. A well designed crystal oscillator circuit will limit the crystal dissipation to a few microwatts.

The second problem concerns phase shift in the amplifier which is used to make the oscillator circuit. For example, the circuit shown in Fig. 7.6 would be a suitable basis upon which to build a quartz crystal oscillator, because it uses a series resonant circuit which could be replaced by a crystal. Let us suppose, however, that the amplifier is not quite ideal and has a small phase lag, at the operating frequency, which changes, because of temperature or power supply fluctuations, by a small amount $\Delta\phi$. This will mean that the frequency of oscillation will have to change as well, in order to keep zero phase shift around the oscillator loop. This shift in frequency is given by $\Delta f = \Delta\phi(f_0/2Q)$, where f_0 is the resonant frequency [16]. Now a $\Delta f/f_0$ of less than 10^{-9} is expected from a good crystal oscillator, so that if Q is 10^5 we shall have to keep $\Delta\phi$ below 2×10^{-4} radians, or about 0.02°. This means that the bandwidth of the amplifier used to make the oscillator loop should be much greater than the crystal frequency.

7.9 An experimental crystal oscillator

For our experimental circuit, the one published by Baxandall some time ago is a good choice as it is a first class design and is also particularly simple.

Fig 7.12 shows a version of the Baxandall crystal oscillator circuit [17] which uses currently available transistors. It is immediately obvious that this circuit is based upon the kind of oscillator shown in Fig. 7.6. The amplifier, A, in Fig. 7.6, has two inputs: one non-inverting and one inverting. In Fig. 7.12 we have an amplifier which has only one input but has two outputs: a non-inverting output at the collector of Q_2 and an inverting output at the emitter of Q_2.

The positive feedback, via the crystal, is taken from only a fraction of the non-inverting output by means of the potential divider $R_6/(R_6 + R_7)$. The level of oscillation in Fig. 7.12 is not controlled by increasing the negative feedback, which is via R_5, as the level builds up (this was the technique used in Fig. 7.6), but by reducing the positive feedback. This is done by means of the two Schottky barrier diodes, D_1 and D_2 in Fig. 7.12, which are connected to the collector of Q_2. The amplitude at the collector of Q_2 is thus limited to a few hundred millivolts peak.

Fig. 7.12.

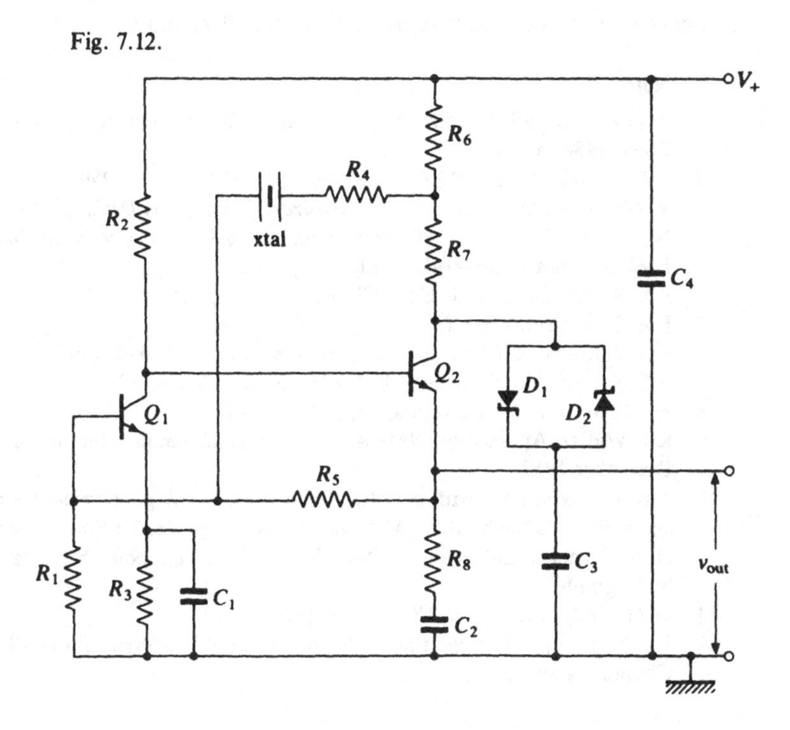

The most important feature of this circuit is that the amplifier, implemented by means of Q_1 and Q_2, should have very wide bandwidth. This is to satisfy the conditions discussed at the end of the previous section concerning the very small phase shift which the amplifier might introduce into the feedback loop. Q_1 and Q_2 are high frequency devices with an f_T well over 500 MHz and are operating at collector currents of a few milliamps. D_1 and D_2 have very low capacitance. This means that the circuit should be suitable for use with any crystal, from quite low frequencies up to perhaps 10 MHz.

The bias conditions are very well defined in this circuit: the base of Q_2 is put at about $V_+/2$, this determines R_2 for whatever I_{C1} has been chosen, and then R_5 and R_1 set the voltage at the base of Q_1. The value of R_3 is thus fixed. The gain of the output stage is then free to be chosen over quite a wide range as it is $(R_6 + R_7)/R_8$, R_8 being kept small. On top of this flexibility we have the potential divider, $R_6/(R_6 + R_7)$, to keep the voltage across the crystal really low compared to the limits set by D_1 and D_2, and also the further possibilities of R_4. R_4 is inserted to actually reduce the Q of the crystal so that we may observe the effects of this upon the waveforms. When the circuit is working correctly, a very good sinusoidal output is obtained from the emitter of Q_2 and the crystal dissipation is only about 1 μW.

The Appendix gives all the constructional details for building Fig. 7.12 and suggestions for the measurements that should be made.

Notes

1 P. Horowitz and W. Hill, *The Art of Electronics*, Cambridge University Press, 1980, p. 166.
2 P.J. Baxandall, *Proc. IEE*, **106B**, *Suppl.* 15–18, 748–58, 1959.
3 V. Manassewitsch, *Frequency Synthesizers: Theory and Design*, J. Wiley, New York, 2nd edn, 1980; W.F. Egan, *Frequency Synthesis by Phase Lock*, J. Wiley, New York, 1981.
4 R.S. Sidorowicz, *Int. J. Circ. Theor. Applic.*, **3**, 135–48, 1975.
5 L.A. Meacham, *Proc. IRE*, **26**, 1278–94, 1938.
6 R.S. Sidorowicz, *Electronic Eng.*, **39**, 498–502 and 560–4, 1967.
7 R.S. Sidorowicz, *Proc. IEE*, **119**, 283–93 and 1283–4, 1972.
8 A. Mayer, *Int. J. Electronics*, **52**, 263–74, 1982.
9 R.J. Widlar, Application Note AN–29, National Semiconductor Corp., December 1969.
10 Also termed a filter with a Butterworth response. A good reference on active filters is the book by M.E. van Valkenburg, *Analog Filter Design*, Holt, Rinehart and Winston, New York, 1982. This book has a good bibliography.
11 W.G. Cady, *Proc. IRE*, **10**, 83–114, 1922.
12 J.F. Nye, *Physical Properties of Crystals*, Oxford University Press, 1957, Chapters 4 and 7.

13 P.J. Baxandall, *Radio and Electronic Eng.*, **29**, 229–46, 1965.
14 Note 13 above, p. 245.
15 Two recent texts which give considerable technical detail on a wide range of crystal oscillator circuits are: R.J. Matthys, *Crystal Oscillator Circuits*, J. Wiley, New York, 1983, and M.E. Frerking, *Crystal Oscillator Design and Temperature Compensation*, Van Nostrand Reinhold Co., New York, 1978.
16 This follows from $Z = R_1 + j\omega L_1[1 - (\omega_0/\omega)^2]$, where $\omega_0 = 2\pi f_0 = 1/\sqrt{L_1 C_1}$. We then write $\tan\phi = \omega L_1[1 - (\omega_0/\omega)^2]/R_1$, $Q = \omega L_1/R_1$ and use $(\omega^2 - \omega_0^2) \approx 2\omega(\omega - \omega_0)$ along with $\phi \approx \tan\phi$ for small ϕ.
17 Note 13 above.

8

Translinear circuits

8.1 Translinearity

Circuits in which the 'primary function arises from the exploitation of the proportionality of transconductance to collector current in bipolar transistors', have been called 'translinear' by Gilbert [1] and this terminology will be adopted here. Translinear circuits can only be realised as integrated circuits. The matching of bipolar transistors in geometry and temperature plays a significant role in translinear circuit design.

The point behind translinear circuits is the fact that the collector current of a bipolar transistor is found, experimentally, to obey the equation

$$I_C = I_{CBO} \exp[V_{BE}/(kT/e)] \quad (8.1)$$

over a very wide range of collector current. Equation (8.1) follows from simple bipolar transistor theory [2] and, in the kind of integrated circuits we shall be considering here, experimental evidence shows that it is valid, in one and the same device, from collector currents of 15 pA up to 1.5 mA, or over five decades [3].

In equation (8.1), I_{CBO} is the saturation reverse bias current of the collector junction, and is a constant at constant temperature. As is well known, when we differentiate equation (8.1) with respect to V_{BE}, and so obtain an expression for the transconductance,

$$g_m = dI_C/dV_{BE} = I_C/(kT/e) \quad (8.2)$$

we find that g_m is proportional to collector current.

Equation (8.2) immediately suggests the possibility of using bipolar transistors to build an amplifier which has a gain that can be varied, linearly, by means of some external control signal. We have already introduced the idea of such a variable gain block in the previous chapter as part of Fig. 7.4. In that example, a dual gate MOST was proposed as a suitable device to give a voltage controlled gain. This is possible, but the gain would then only be controllable over a small range and linearity is certainly not expected from MOS device theory, whereas bipolar device theory predicts the possibility of linear gain control over a very wide range indeed.

As an example of such an amplifier with variable gain, consider Fig. 8.1. Here Q_1 and Q_2 form the simple current mirror which has been covered here in Chapter 3, Fig. 3.6. Q_3 and Q_4 form a balanced long tailed pair amplifier, with equal collector load resistors both labelled R_2. This circuit was also a topic of Chapter 3: Fig. 3.4(*a*).

In Fig. 8.1, the control voltage input, V_{cont}, sets the current in the current mirror so that

$$I_{C2} = (V_{\text{cont}} - V_{BE1} + |V_-|)/R_1 \tag{8.3}$$

and we may assume, as this is an integrated circuit, that I_{C2} is shared equally by Q_3 and Q_4 when $V_{\text{in}} = 0$.

When an input voltage is applied to Fig. 8.1, we can use equation (8.1), along with

$$V_{\text{out}} = (I_{C3} - I_{C4})R_2 \tag{8.4}$$

$$I_{C2} = (I_{C3} + I_{C4}) \tag{8.5}$$

and the identity,

$$\tanh(x) = (e^x - e^{-x})/(e^x + e^{-x}) \tag{8.6}$$

Fig. 8.1.

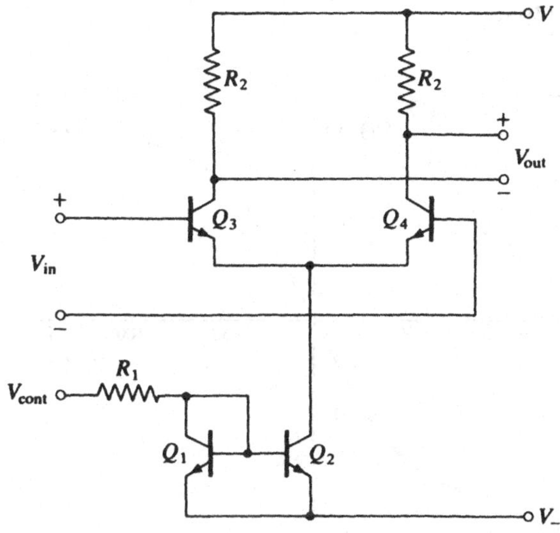

to obtain the concise expression

$$V_{out} = I_{C2}R_2 \tanh[V_{in}/2(kT/e)] \tag{8.7}$$

for the transfer function of this circuit.

Equation (8.7) is sketched in Fig. 8.2 for the quite arbitrarily chosen values, $I_{C2} = 1\,\text{mA}$ and $R_2 = 1\,\text{k}$. It has been assumed that (kT/e) has a room temperature value of 25 mV. As Fig. 8.2 shows, the circuit of Fig. 8.1 has a reasonably constant gain, dV_{out}/dV_{in}, for input voltages below about 20 mV, and that this gain is about 20, which is, as expected, the value of $g_m R_2$ where g_m is given by equation (8.2). The amplifier saturates for inputs greater than about 100 mV in magnitude, but the small signal gain should be proportional to I_{C2}, and thus to V_{cont} in view of equation (8.3), over a very wide range.

8.2 The operational transconductance amplifier (OTA)

Fig. 8.1 is an academic circuit and any attempt to make a useful amplifier from this beginning takes us back to Chapter 3 and the problem of level shifting, discussed in Sections 3.3 and 3.7. In Fig. 8.1 we do not only have the problem that the two output terminals must both be several volts above the level of the two input terminals, but we also have to remember that this shift in level is not constant in this circuit. As V_{cont} is reduced, to reduce the gain, the voltage level at both output terminals rises towards V_+. For a useful variable gain amplifier we should like to have the output terminals always at zero mean level. We should also probably prefer a single-ended output.

Fig. 8.2.

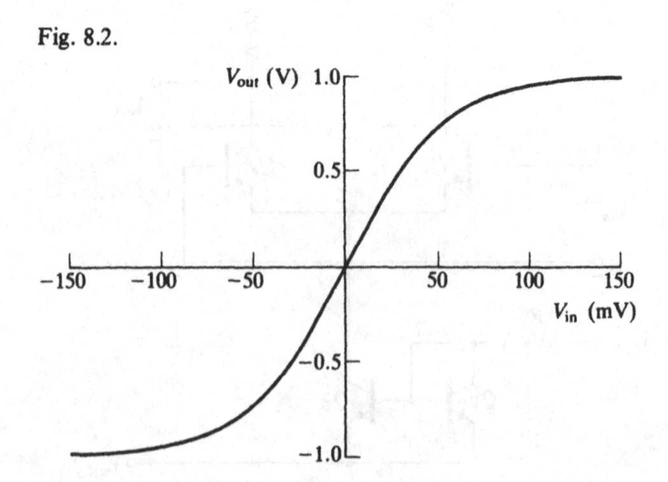

A very elegant solution to this problem was published in 1969 by Wheatley and Wittlinger [4] as a device they named an 'operational transconductance amplifier' or OTA. A simplified version of the OTA circuit is shown in Fig. 8.3. Here again we have the input arrangement of Fig. 8.1: a long tailed pair balanced input amplifier with a current mirror, Q_1 and Q_2, as its long tail. The tail current is now controlled by an external control current, I_{cont}, instead of the voltage control of Fig. 8.1. Fig. 8.3 represents an integrated circuit and, in what follows, we shall assume that all the transistors have identical geometry and all operate isothermally.

In place of the two identical load resistors, both labelled R_2 in Fig. 8.1, the OTA of Fig. 8.3 has two identical current mirrors: Q_5, Q_6 and Q_8, Q_9. These mirror I_{C3} and I_{C4} as I_{C6} and I_{C9} respectively, but note that I_{C6} and I_{C9} are now sources from the V_+ voltage level. The collectors of Q_3 and Q_4, in contrast, were sinking current.

I_{C6}, in Fig. 8.3, is taken to another current mirror in the circuit: Q_7, Q_{10}. This again changes I_{C6} back into a sink of current, but now sinking directly to the V_- level in the circuit. We may now connect I_{C9}, a source, to I_{C10}, a sink, and obtain an output terminal, I_{out}, which will supply

$$I_{out} = (I_{C4} - I_{C3}) \tag{8.8}$$

simply because I_{C4} has been mirrored to I_{C9} while I_{C3} has been mirrored, via I_{C6}, to I_{C10}. This output terminal has the remarkable property of being able to supply the same current, $(I_{C4} - I_{C3})$, regardless of the voltage level it may be connected to, provided this voltage level lies in between V_+ and V_-.

In Fig. 8.3 we have a simple identity between I_{cont} and I_{C2}, in place of our previous equation (8.3). Equation (8.5), of the previous analysis, is now

$$I_{cont} = (I_{C4} + I_{C3}) \tag{8.9}$$

Fig. 8.3.

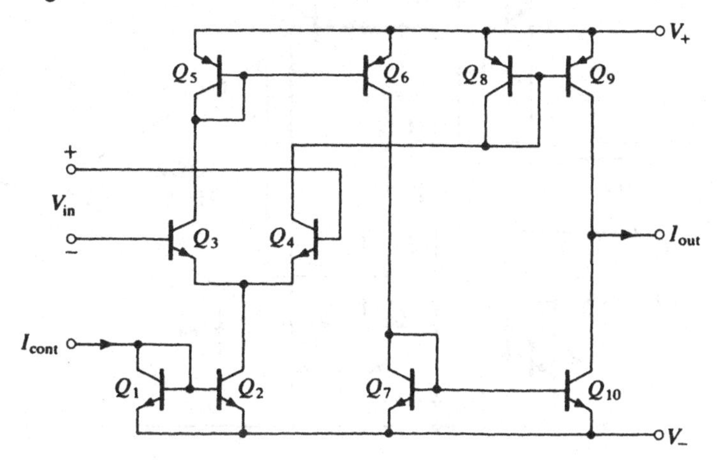

so that we can again use equations (8.1) and (8.6) to obtain

$$I_{out} = I_{cont} \tanh[V_{in}/2(kT/e)] \quad (8.10)$$

as the transfer function of the OTA. This is, of course, exactly the same shape as the function sketched in Fig. 8.2 and, for small values of V_{in}, is a transconductance

$$dI_{out}/dV_{in} = I_{cont}/2(kT/e) \quad (8.11)$$

Referring back to equation (8.7), it is clear that the OTA can be turned into a voltage amplifier, simply by connecting a load resistor, R_2, to the output terminal of Fig. 8.3. The small signal gain may now be controlled by varying I_{cont} and the d.c. level at the output terminal will not vary. Furthermore, the d.c. level at the two input terminals of Fig. 8.3 may be at any value between (V_+ − 700 mV) and (V_- + 1.4 V). The transistors Q_3 and Q_4, for small signal input levels, will still be active. Altogether, Fig. 8.3 is a remarkable new circuit shape of considerable potential [5].

8.3 Linearisation of translinearity

An interesting improvement to the OTA, described in the previous section, is shown in Fig. 8.4. Here, two transistors, Q_{15} and Q_{16}, are connected across the input terminals of the original circuit, shown in Fig. 8.3. Both Q_{15} and Q_{16} are clamped active, by simply connecting base and collector together, and are both forced to pass a current, I_D, which is set externally via the current mirrors: Q_{11}, Q_{13}, Q_{17} and Q_{19}; Q_{12}, Q_{14} and Q_{18}.

Fig. 8.4.

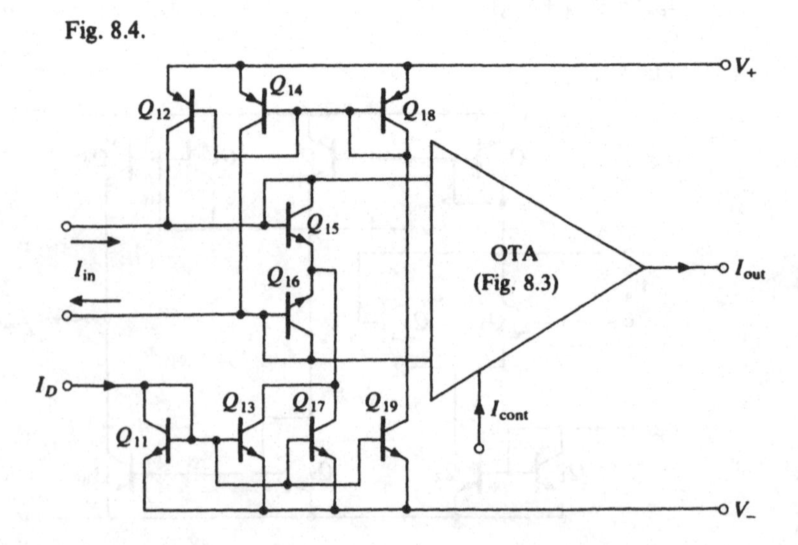

The idea behind Fig. 8.4 is to remove the non-linearity from the OTA transfer function, the curvature shown in Fig. 8.2, and also to eliminate the temperature dependence which is evident from equation (8.10). The origin of this idea will be found in an important paper by Gilbert [6] and it has been discussed in detail, from the point of view of the OTA, by Kaplan and Wittlinger [7].

We can see how the new idea works by writing equation (8.1) as

$$V_{BE} = (kT/e)\log_e(I_C/I_{CBO}) \tag{8.12}$$

Equation (8.12) may now be used to express the fact that the sum of the base–emitter voltages of Q_{15} and Q_{16} must always equal the sum of the base–emitter voltages of the two input transistors of the OTA, Q_3 and Q_4 in Fig. 8.3. As the collector currents of Q_{15} and Q_{16} are $(I_D + I_{in})$ and $(I_D - I_{in})$ respectively, this means, cancelling out the common term (kT/e) and assuming that all four transistors have the same collector saturation current, I_{CBO}, that

$$(I_D + I_{in})/(I_D - I_{in}) = I_{C3}/I_{C4} \tag{8.13}$$

Now equations (8.8) and (8.9) may be used to write equation (8.13) as

$$I_{out} = I_{in}(I_{cont}/I_D) \tag{8.14}$$

Equation (8.14) shows a simple, temperature independent, relationship between the input and output currents of the new circuit, Fig. 8.4. The constant connecting I_{in} and I_{out} is (I_{cont}/I_D), and this may be varied over the same wide range as before by varying I_{cont}. Control is also possible by varying I_D, but this is restricted by the fact that I_D must always be greater than I_{in}. Experimentally, the linearity between I_{out} and I_{in} is found to be excellent over the range $-I_D < I_{in} < +I_D$ [8].

8.4 An experimental circuit

As an example of an application of an OTA, and as the experimental circuit for this chapter, let us consider the possibility of building a very non-linear electrical resonator, or filter, which will show some of the remarkable jump phenomena and sub-harmonic resonances peculiar to such non-linear systems [9].

A simple *linear* resonator may be imagined as a mass, m, hanging on the end of a spring which produces a restoring force, Kx, where x is the displacement of the mass away from the equilibrium position that it would normally take up due to gravity. The differential equation of such a system is

$$m\mathrm{d}^2x/\mathrm{d}t^2 + Kx = 0 \tag{8.15}$$

and this equation tells us that the system oscillates at a constant frequency,

$\omega = \sqrt{K/m}$, and at whatever constant amplitude might be set by the initial conditions.

Let us consider the possibility of building a resonator with a very *non-linear* kind of spring. In fact, let us choose a really non-linear function for the restoring force: like the one sketched in Fig. 8.5. Here, y represents the force produced by the spring, as it is extended or compressed by a distance x. It is clear that we are now dealing with what is usually termed a hard-spring: the restoring force is very weak for small displacements, but it increases far more rapidly than the linear spring as the displacement builds up. The particular function sketched in Fig. 8.5 is

$$y = Kx|x| \tag{8.16}$$

where K has been chosen equal to unity. If we can make an electrical network which has equation (8.16) as its transfer function, then it is a simple matter to build an electrical analog of the non-linear resonator which has the differential equation

$$m\mathrm{d}^2x/\mathrm{d}t^2 + Kx|x| = 0 \tag{8.17}$$

Fig. 8.6 gives the circuit diagram of a system that will give a transfer function of the form given by equation (8.16) for *positive* values of x only. Negative values of x are dealt with by a very similar circuit which will be considered below.

Fig. 8.6 uses the CA3280 device, which is an integrated circuit containing two of the linearised OTA devices shown in Fig. 8.4. Because Fig. 8.6 is part

Fig. 8.5.

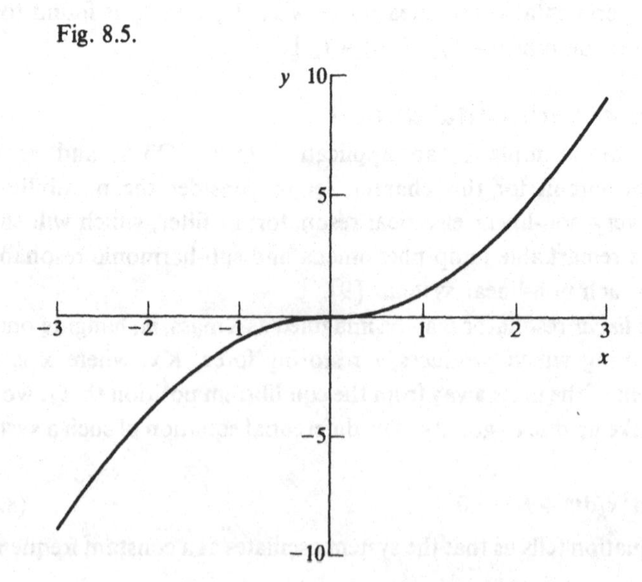

of the experimental circuit for this chapter, the pin numbers relating to the CA3280 have been added.

The first thing to note about Fig. 8.6 is that it uses four identical resistors, two for each half of the CA3280, all labelled R_1, to convert an input voltage, V_{in}, into an input current, I_{in}. Because the voltage across the input terminals of the CA3280 will never exceed about 50 mV, in any sensible application, we may assume that

$$I_{in} = V_{in}/2R_1 \tag{8.18}$$

The second thing to note about Fig. 8.6 is that an operational amplifier is used to convert the output current of the top half of the CA3280 into a voltage. Taking note of the signs in Fig. 8.6, we may now combine equations (8.18) and (8.14) to give

$$V_{out} = V_{in}(R_5/2R_1)(I_{cont\,1}/I_D) \tag{8.19}$$

Both halves of the CA3280 are arranged to have the same current, I_D. The current I_{cont}, however, is varied in the top half because it is the output current of the lower half of the device. This lower half has a constant value of I_{cont}, $I_{cont\,2}$, and it follows from equations (8.14) and (8.18) that

$$I_{cont\,1} = (V_{in}/2R_1)(I_{cont\,2}/I_D) \tag{8.20}$$

Fig. 8.6.

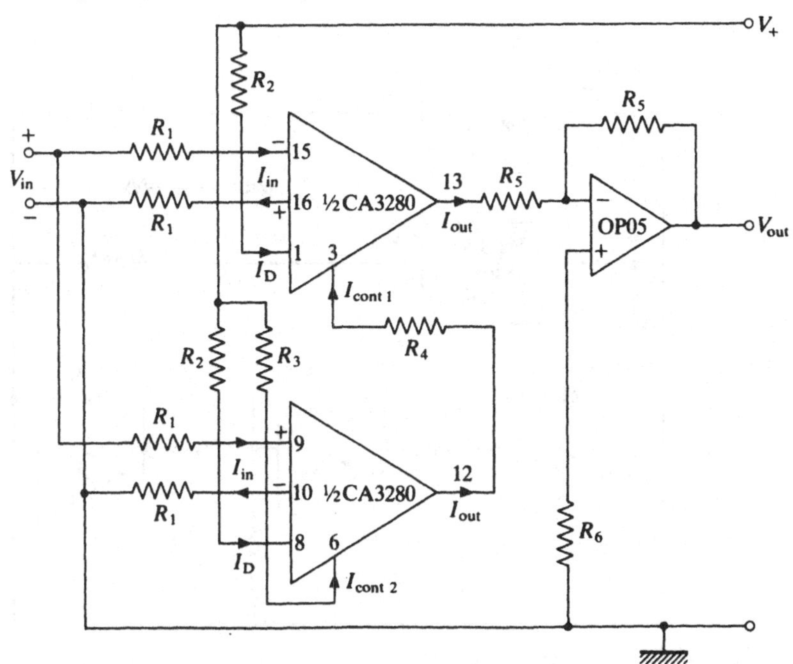

Combining equations (8.19) and (8.20) then gives the relationship needed for the positive quadrant of Fig. 8.5:

$$V_{out} = V_{in}^2(I_{cont\,2}R_5)/(2I_DR_1)^2 \tag{8.21}$$

When V_{in} is made negative, the output from the circuit shown in Fig. 8.6 is simply zero because the lower half of the CA3280 tries to make $I_{cont\,1}$ negative and this will turn off the upper half of the CA3280 completely. The way to handle the negative quadrant of the function shown in Fig. 8.5 is to build another circuit, identical to the one shown in Fig. 8.6, but with the leads to the input terminals of the lower half of the CA3280 reversed.

This is shown in Fig. 8.7 where the experimental circuit is given for the complete analog of the hard-spring resonator described by equation (8.17). If we write equation (8.21) as

$$V_{out} = KV_{in}^2 \tag{8.22}$$

the analog shown in Fig. 8.7 has the equation of motion

$$(C_7R_7)^2d^2v/dt^2 + Kv|v| = -V\sin(\omega t) \tag{8.23}$$

where v is the voltage at the output of A_1 and $V\sin(\omega t)$ is applied to the input terminal. For experimental work, it is necessary to add some damping to this dynamic system, and this is easily done by putting a resistor in parallel with the capacitor which is across A_2. This is discussed in

Fig. 8.7.

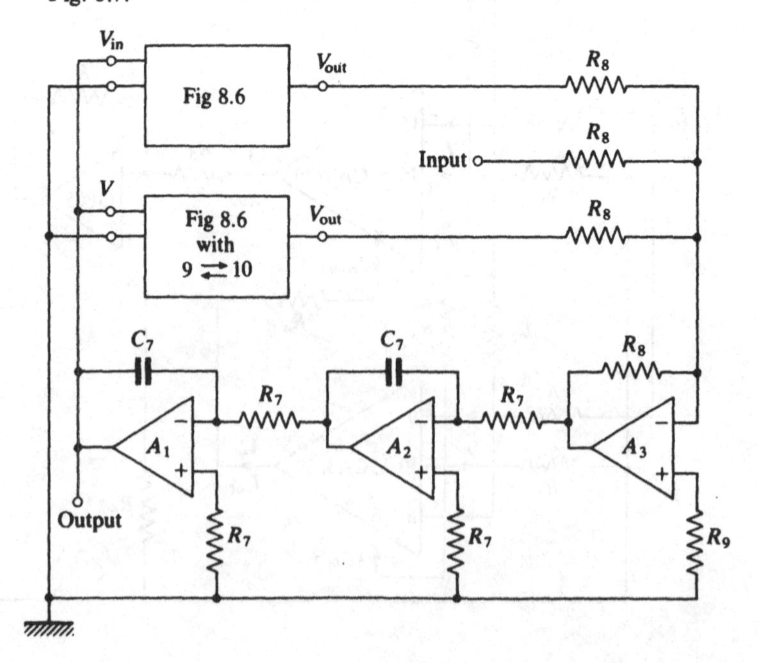

detail in the Appendix along with other small modifications which need to be made before the system shown in Fig. 8.7 becomes a practical experimental circuit. The behaviour of this highly non-linear dynamic system is quite remarkable and is just as interesting as the electronic circuit design problems that it involves.

8.5 Further applications of the OTA

Operational transconductance amplifiers, both the linearised variety, discussed in the previous section, and the simpler version of Fig. 8.3, are useful in a wide variety of applications.

An excellent sample-and-hold system using the CA3080 OTA has been published [10]. This uses a circuit of the form of Fig. 8.3 as a gating amplifier to charge up a sampling capacitor. The form of the output circuit shown in Fig. 8.3 is obviously ideal for this because, when $I_{cont} = 0$, which puts both Q_9 and Q_{10} off, the output terminal looks like a very high impedance indeed. Also, for sample-and-hold applications, the OTA works at its maximum I_{cont} because maximum output current is needed for high speed charging of the sampling capacitor. The maximum value of I_{cont} also gives the OTA its maximum bandwidth.

Applications as variable gain amplifiers in audio systems and in voltage controlled filters are fairly obvious tasks for the OTA. In the previous section, one OTA was made to control a second OTA so that a squaring circuit was implemented. A simple extension of this idea, where different signals are supplied to the two OTA inputs, will produce an analog multiplier which will operate in two (x, y) quadrants [11], and four quadrant multiplication is also possible [12], although this is better done with a related translinear device which is specially designed for the purpose [13].

8.6 Absolute temperature measurement and voltage references

Another application of the translinear idea involves the fact that the difference between the base–emitter voltages of two transistors, carrying different collector currents but identical in all other respects, varies linearly with absolute temperature.

Consider two identical transistors, Q_1 and Q_2, side by side in a monolithic silicon device so that they may be assumed to be at the same temperature, T. Let Q_1 have a collector current I_{C1} and Q_2 have a collector current I_{C2}. Then, as we may assume that Q_1 and Q_2 have the same reverse collector saturation current, I_{CBO}, we have

$$V_{BE1} = (kT/e)\log_e(I_{C1}/I_{CBO}) \tag{8.24}$$

and

$$V_{BE2} = (kT/e)\log_e(I_{C2}/I_{CBO}) \qquad (8.25)$$

from equation (8.12). It follows that

$$\Delta V_{BE} = V_{BE1} - V_{BE2} = (kT/e)\log_e(I_{C1}/I_{C2}) \qquad (8.26)$$

Equation (8.26) is the basis of two useful electronic devices. The first is an absolute temperature sensor, which consists of a very accurately matched pair of transistors fed with collector currents, I_{C1} and I_{C2}, in some convenient ratio. The difference voltage, ΔV_{BE}, is made the input of a differential amplifier, perhaps one of the kind discussed in Chapter 4, Fig. 4.3, and the output from this amplifier gives a measure of the absolute temperature, T. Such an electronic thermometer involves no really new circuit shapes or problems [14] and will not be discussed in detail here.

An aspect of equation (8.26) which does lead us to some interesting and new circuit shapes, is the idea of generating a temperature independent voltage reference by combining the linearly increasing voltage, given by equation (8.26), with some other available voltage which falls with increasing temperature. The most familiar voltage, in solid state electronics, which falls with an increase in temperature is the V_{BE} of an active bipolar transistor: if the collector current is to be constant, the V_{BE} will have to be reduced by about 2.1 mV/°C.

The positive temperature coefficient of ΔV_{BE} can be found from equation (8.26) and is much smaller than 2.1 mV/°C. In order to make a temperature independent voltage reference, we need to amplify ΔV_{BE}, given by equation (8.26), and add this to the V_{BE} of one of the two matched transistors. This is a very interesting problem in inventing new circuit shapes, and the solutions take many forms. The idea was first published some time ago [15], and really excellent voltage references using this principle are now available [16].

One example of such a voltage reference will be given here. This is shown in Fig. 8.8. The difference between the base–emitter voltage of Q_1 and Q_2 is, very approximately, amplified by R_3/R_1 while the absolute value of V_{BE} is added to this amplified voltage, thus forming V_{out} as the temperature independent reference. This is a summary of the reasoning behind the circuit shape of Fig. 8.8. Some simple analysis is needed to see the detail.

If we assume that the operational amplifier, A, in Fig. 8.8, is ideal, then

$$I_{C1}R_1 + (kT/e)\log_e(I_{C1}/I_{CBO}) = (kT/e)\log_e(I_{C2}/I_{CBO}) \qquad (8.27)$$

follows when we equate the true input voltage of the amplifier to zero. Similarly, assuming zero input bias current, we must have

$$I_{C1}R_3 = I_{C2}R_2 \qquad (8.28)$$

and equations (8.27) and (8.28) then give a solution for I_{C1},

$$I_{C1} = (kT/eR_1)\log_e(R_3/R_2) \tag{8.29}$$

Now the output voltage may be written

$$V_{out} = I_{C1}(R_1 + R_3) + (kT/e)\log_e(I_{C1}/I_{CBO}) \tag{8.30}$$

and differentiated with respect to temperature to give

$$\mathrm{d}V_{out}/\mathrm{d}T = (k/e)[1 + (1 + R_3/R_1)\log_e(R_3/R_2)] - (E_g - V_{BE1})/T \tag{8.31}$$

where we have used equations (8.29) and (8.12), together with the approximation

$$I_{CBO} = I_0 \exp(-E_g/(kT/e)) \tag{8.32}$$

to express the strong temperature dependence of I_{CBO}. In equation (8.32), I_0 is treated as a constant and E_g is the band-gap for silicon: 1.1 V at room temperature. Another simplifying assumption in equation (8.31) is that V_{BE1} is being treated as a constant once $\mathrm{d}V_{out}/\mathrm{d}T$ has been formed. These approximations are avoided in a full analysis of the problem [17], but here we need a concise way of understanding how the circuit works.

Equation (8.31) shows that V_{out}, the reference voltage generated by the circuit shown in Fig. 8.8, should be independent of temperature if we make

$$1 + (1 + R_3/R_1)\log_e(R_3/R_2) = (E_g - V_{BE1})/(kT/e) \tag{8.33}$$

Suppose we choose to make $R_3/R_2 = 4.7$, quite arbitrarily, and argue that, at room temperature, $E_g = 1.1$ V, $V_{BE1} = 650$ mV and $(kT/e) = 25$ mV. Equation (8.33) then tells us that the gain, $(1 + R_3/R_1)$, should be 11.0. Suitable values for R_1, R_2 and R_3 may now be chosen: for example, $R_1 = 470\,\Omega$, $R_2 = 1$ k and $R_3 = 4.7$ k. With these values, the output voltage follows from equations (8.29), (8.30) and (8.12). The result is 1.076 V, which is some 10% below the exact solution, although very close

Fig. 8.8.

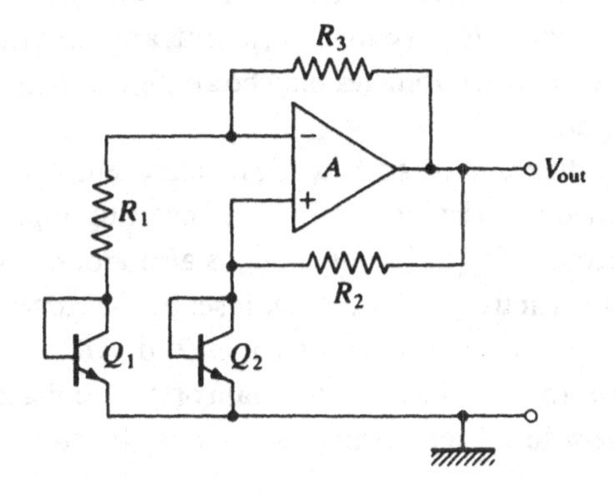

to the value of $E_g = 1.1$ V which we have assumed here, and have also assumed to be independent of temperature. As would be expected, from elementary considerations, the temperature independent value of V_{out} should equal the band-gap for silicon at 0 °K, because this is the value V_{BE} extrapolates back to as $T \rightarrow 0$, while $\Delta V_{BE} \rightarrow 0$ as $T \rightarrow 0$. Our voltage reference is based upon the sum of V_{BE} and a multiple of ΔV_{BE}. It is for this reason that these voltage reference circuits are often called 'band-gap references'.

8.7 The logarithmic amplifier

There are at least two quite distinct interpretations of the term 'logarithmic amplifier' in electronic circuit design. The most common interpretation is the kind of amplifier found as the IF amplifier in a radar system. This is a wide bandwidth, high frequency, amplifier which must handle input signals over a very wide range of amplitudes. We shall not consider this kind of logarithmic amplifier here. The second interpretation of the term is a direct coupled instrumentation amplifier which must handle input voltages or currents varying over a range of several decades, and give an output level which is suitable for a simple analog indicating instrument. A typical application would be the measurement of the neutron flux in a power reactor as the reactor was taken up to full output.

Such a logarithmic instrumentation amplifier could be built by taking an operational amplifier and connecting the output back to the inverting input via a bipolar transistor. As equation (8.12) shows, current fed back in this way, to the virtual earth of the amplifier, should produce a change in output voltage proportional to the logarithm of any other current which is also fed into the virtual earth. Several interesting circuit shapes have been proposed to implement this idea, and a useful review has been given by Dobkin [18]. The use of a single bipolar transistor, as outlined above, is not satisfactory because of the offset voltage that this involves. A second transistor should be brought in, operating at constant current, and then the difference between the two base–emitter voltages may be amplified to give the required logarithmic output.

Fig. 8.9 shows the circuit diagram of such a logarithmic amplifier. Operational amplifier A_1 provides a virtual earth, at its inverting input, which is connected to the collector of Q_1. The base of Q_1 is also grounded, and this means that Q_1 is clamped active. The V_{BE} of Q_1 is set by the current in R_1, V_{in}/R_1, because this becomes the collector current of Q_1 down to the output of A_1. Resistor R_2 is inserted to reduce the loop gain of this feedback path: stabilisation is a problem in all logarithmic amplifiers. It follows, from equation (8.12), that

$$V_{BE1} = (kT/e)\log_e[(V_{in}/R_1)/I_{CBO1}] \tag{8.34}$$

Transistor Q_2, in Fig. 8.9, carries a constant current: approximately V_+/R_5 in this circuit, and could be made more precise at the expense of very little complication. It follows that

$$V_{BE2} = (kT/e)\log_e[(V_+/R_5)/I_{CBO2}] \tag{8.35}$$

Now Q_1 and Q_2 will be the two halves of a very accurately matched pair of npn transistors [19]. This means that, in equations (8.34) and (8.35), the temperatures may be assumed equal, and thus $I_{CBO1} = I_{CBO2}$. The input voltage to amplifier A_2, in Fig. 8.9, is then given by $V_{BE2} - V_{BE1}$ and, as A_2 is connected as a feedback amplifier with a gain of $+(R_4 + R_3)/R_3$, it follows that the output voltage from the circuit of Fig. 8.9 is

$$V_{out} = -[(kT/e)(R_4 + R_3)/R_3]\log_e(V_{in}R_5/V_+R_1) \tag{8.36}$$

The output voltage is zero when $V_{in} = V_+(R_1/R_5)$, and varies linearly with the logarithm of V_{in} as V_{in} goes above or below this level. The circuit will only work correctly if the temperature is kept constant, but this problem may be overcome by making either R_3 or R_4 temperature dependent so that the term (kT/e), which occurs in equation (8.36), is compensated for. Another problem with this circuit is that very small input bias currents are called for in the operational amplifiers, A_1 and A_2, so that devices with input bias current cancellation are needed: the kind of operational amplifier discussed in Section 3.8. A useful experimental circuit along the lines of Fig. 8.9 has been described by Nicholson and Miller [20], who give details of the measurements they made on the performance: five decades of input voltage variation could be accommodated quite easily.

Fig. 8.9.

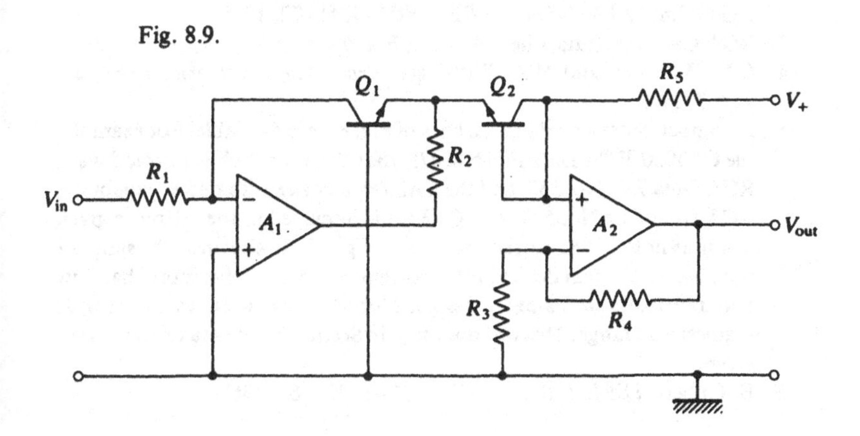

8.8 Trigonometric functions

This chapter ends with a brief mention of some recent work in translinear circuits which makes the generation of trigonometric functions possible. This idea is based upon the function shown in Fig. 8.2: $y = \tanh(x)$.

By using long tailed pair circuits, along the lines of Fig. 8.1, it is possible to generate a whole series of functions of the form $y = \pm\tanh(x \pm na)$. These look identical to the function of Fig. 8.2, displaced to left or right along the x axis by amounts a, $2a$, $3a$ and so on, and with either positive going or negative going slope.

The idea of adding such functions together to generate a few cycles of a function that looks like a sinusoid was proposed by Gilbert [21] in 1977, and he pointed out the unexpected fact that a judicious choice of the shift parameter, a, would allow a very accurate synthesis of the function $A\sin(x)$, over a limited range of x. Just how limited the range of x was, depended upon the number of long tailed pairs used.

In a later paper [22], Gilbert extended these ideas and described circuits which could synthesise functions of the form $A\sin(u - v)/\sin(x - y)$, where u, v, x and y may all be variable inputs to the circuit. Such flexibility makes the generation of virtually any trigonometric function possible, provided always that it is bounded. This most recent development in the field of translinear circuits should suggest all kinds of interesting experimental work for circuit designers who have access to an integrated circuit process.

Notes

1 B. Gilbert, *Electronics Letters*, **11**, 14–16 and 136, 1975.

2 Although we must be careful when dealing with bipolar transistors which are intended for very high frequency work. An important review of transistor modelling which covers this is the paper by P. Rohr and F.A. Lindholm, *IEEE J. Sol. St. Circ.*, **SC–10**, 65–72, 1975.

3 RCA CA3280: Data File No. 1174, Fig. 6.

4 C.F. Wheatley and H.A. Wittlinger, *Proc. Nat. Electronics Conf.*, **25**, 152–7, 1969.

5 A number of devices along the lines of Fig. 8.3 are available. For example: the CA3080, RCA Data File No. 475; the CA3060, which is a triple device, RCA Data File No. 537; and the CA3094, a device with power capability, RCA Data File No. 598. The CA3094 is discussed in the reference given below as note 7. The current mirrors in Fig. 8.3 are shown as the simplest possible. In the real devices these current mirrors involve more than just two transistors each and are designed for very high accuracy over a wide temperature range. This was discussed in Section 3.7, where references are given.

6 B. Gilbert, *IEEE J. Sol. St. Circ.*, **SC–3**, 353–65, 1968.

7 L. Kaplan and H.A. Wittlinger, *IEEE Trans. Broadcast and Television Receivers*, **BTR–18**, 164–75, 1972.
8 RCA CA3280: Data File No. 1174, Fig. 12(*a*).
9 N. Minorsky, *Non-linear Oscillations*, Van Nostrand, Princeton, 1962.
10 The design is given as part of the data sheet for the CA3140 operational amplifier: RCA Data File No. 957, Fig. 34.
11 H.A. Wittlinger, RCA Appl. Note ICAN–6668.
12 H.A. Wittlinger and D. Nissman, RCA Appl. Note ICAN–6818.
13 B. Gilbert, *IEEE J. Sol. St. Circ.*, **SC–3**, 365–73, 1968.
14 A very well designed absolute temperature sensor system is described by J. Simmons and D. Soderquist, 'Temperature measurement method based on matched transistor pair requires no reference'. Precision Monolithics Inc. Appl. Note No. 12, 1981. The design has ±1 °K accuracy over the useful range of 218 °K to 398 °K.
15 US Pat. No. 3617859 of 2 November 1971, to R.C. Dobkin and R.J. Widlar with National Semiconductor Corp.
16 An integrated circuit which generates a voltage reference which is constant to 0.5 ppm/°C over the range −25 °C to +85 °C has been described by G.C.M. Meijer, P.C. Schmale and K. van Zalinge, *IEEE J. Sol. St. Circ.*, **SC–17**, 1139–43, 1982.
17 R.J. Widlar, *IEEE J. Sol. St. Circ.*, **SC–6**, 2–7, 1971; K.E. Kuijk, *IEEE J. Sol. St. Circ.*, **SC–8**, 222–6, 1973.
18 R.C. Dobkin, Appl. Note AN–30, National Semiconductor Corp., December 1969.
19 For example, the MAT-01 manufactured by Precision Monolithics Inc.
20 P.F. Nicholson and S. Miller, *The Bifet Design Manual*, Texas Instruments, Bedford, England, no date.
21 B. Gilbert, *Electronics Letters*, **13**, 506–8, 1977.
22 B. Gilbert, *IEEE J. Sol. St. Circ.*, **SC–17**, 1179–91, 1982.

9

Power amplifiers

9.1 Power gain

The importance of power gain in electronic circuit design has been emphasised in previous chapters: in Chapter 2, in connection with small signal tuned amplifiers, and again in Chapter 5, in connection with a similar small signal problem involving a photodiode. In both these examples the point was that the input signal carried some kind of information, and this very low level input signal had to be brought up to a more usable level where it could be processed.

In this chapter we consider circuits which take the process of power amplification even further. The input signal is now looked upon as carrying the information which is needed to control a flow of power from some power supply to some load where it will perform useful work. A block diagram defining the power amplifier problem in general is given in Fig. 9.1.

The *amplification*, or power gain, concept belongs to the path running from the control signal input, via the power amplifier, to the load. The path running from the power supply to the load may have to involve a considerable *loss* of power simply because of some quite separate requirement of the complete system; for example, bandwidth or linearity.

Fig. 9.1.

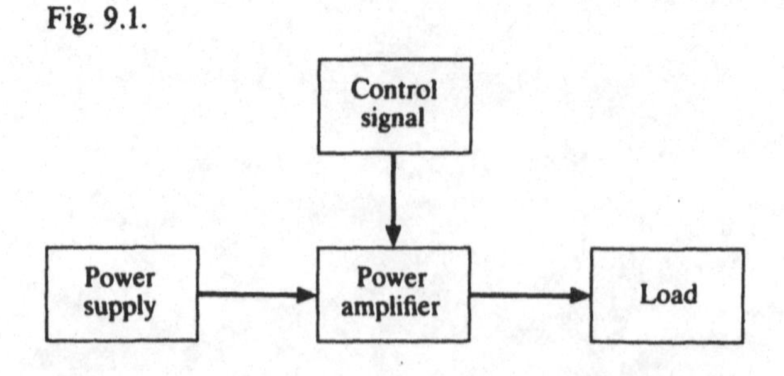

From the general viewpoint of Fig. 9.1, a number of electronic systems may be considered. In this chapter only three will be dealt with. The first might be called the stabilised power supply problem, where the power supply of Fig. 9.1 is a crude source of d.c. or a.c. power, subject to noise and fluctuations, and the control signal demands some special requirement for the power delivered to the load: for example, constant voltage. The second kind of power amplifier problem we shall look at is the audio power amplifier where the power supply is a simple d.c. one, the control signal is information carried by speech or music and the load is a loudspeaker. Finally, the most difficult problem in power electronics will be looked at briefly: the problem of radio frequency power amplification. The technical chapters in this book thus come to an end with the output circuits of a radio transmitter, whereas they began with the input circuits of a radio receiver.

9.2 The classical stabilised power supply

The vast majority of electronic circuit design ideas begin with the assumption that positive and negative voltage supply rails are available. The classical solution to the problem of providing these stabilised voltage levels is to use a circuit of the shape shown in Fig. 9.2, where the practical details have been included because this is the first experimental circuit for this chapter.

In Fig. 9.2, the a.c. supply is transformed, rectified and crudely filtered, to give a d.c. supply across C_1 of about 30 V on no load. This 30 V supply is used to power an amplifier which is designed to reject noise and fluctuations on its power supply rails and only to reproduce a multiple of its input control voltage. In this example the control voltage is produced by a Zener diode, D_1, and will be about 12 V.

Fig. 9.2.

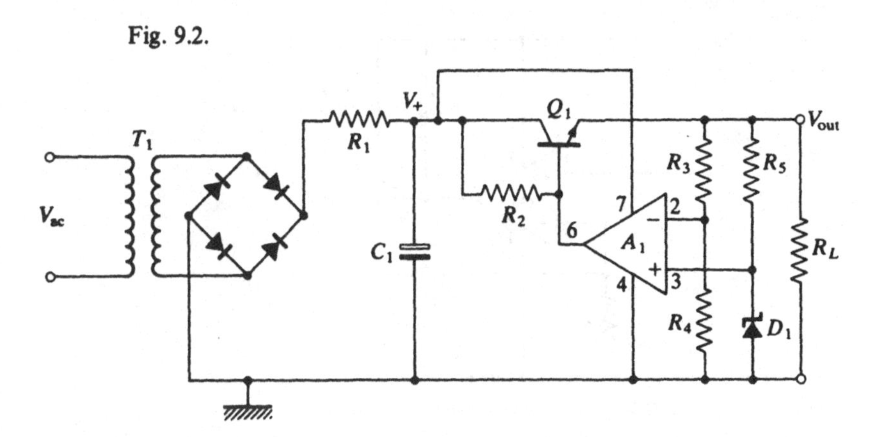

From an academic point of view, Fig. 9.2 could be summarised as 'a simple feedback amplifier'. It is valuable to consider this for a moment because it brings up some important facts which have been emphasised by the particular way Fig. 9.2 has been drawn.

Fig. 9.3 shows a redrawing of Fig. 9.2 as an academic feedback amplifier problem. This tells us straight away that $V_{out} = V_z(R_3 + R_4)/R_4$, whereas this result might only follow after a second look at Fig. 9.2. However, Fig. 9.3 avoids several problems which are particular to the classical stabilised power supply and which must be considered at the beginning. Let us list some of the more obvious:

(1) The original power supply source in Fig. 9.3 is dismissed as 'V_+'. In fact, V_+ has a very large a.c. component, or ripple, and both this and the d.c. level of 'V_+' vary with the load.

(2) The real variable, in both circuits, is R_L. This is not shown in Fig. 9.3.

(3) The control voltage, V_z, is shown in Fig. 9.2 as being derived from the output, that is the stabilised voltage in the circuit. Does this introduce any problems when we ask how the circuit works after it is first switched on? It would obviously not be a good idea to supply the Zener diode from the fluctuating supply, V_+.

These problems make the simple circuit of Fig. 9.2 a very valuable one to experiment with and further practical details are given in the Appendix. It is important to choose an operational amplifier which has an output circuit that will work into the very low impedance presented by the base of Q_1. The operational amplifier must also have good power supply rejection ratio. The resistor R_2 plays an important role in ensuring that the circuit turns on correctly as it supplies current to the emitter junction of Q_1 and then, before

Fig. 9.3.

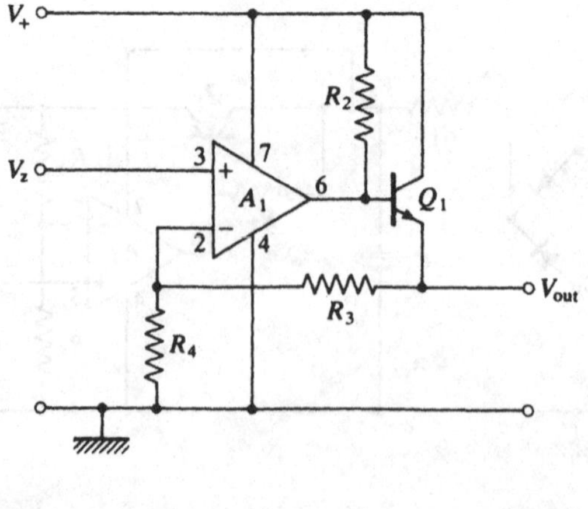

D_1 begins to conduct, to the non-inverting input of A_1 thus causing an initial rise in the output voltage of A_1. It would thus be a mistake to put a capacitor across D_1, because this would spoil this feature of the circuit. Measurements of regulation, ripple and noise at the output, and, most interesting of all, the transient response of the circuit when a load is suddenly applied or removed, are all discussed in the Appendix.

9.3 Switched mode power supplies

One outstanding disadvantage of the classical stabilised power supply is its low efficiency. Referring to Figs. 9.1 and 9.2, power flow from the power supply to the load takes place via the *series regulator*, Q_1 in Fig. 9.2. If the crude power supply which is available across C_1 is fluctuating by many volts, or if the expected changes in d.c. level across C_1, due to load changes, will also be many volts, then several volts drop must always be allowed across Q_1 and the power dissipated in Q_1 may well exceed the power delivered to the load, R_L.

The development of transistors, both bipolar power transistors and MOS power transistors, with excellent switching characteristics, at speeds in the kHz region and at power levels up to several hundred watts, has led to a new kind of stabilised power supply: the switched mode power supply [1]. This can take many forms, but a simple example is shown in Fig. 9.4 which will introduce the concept.

In Fig. 9.4, the same crude d.c. power supply up to C_1 is used as in Fig. 9.2. Let us suppose that the circuit is turned on for the first time. As C_3 is discharged, the reference voltage, V_{ref}, across the Zener diode, D_1, is applied to the non-inverting input of A_1 with little attenuation because, as we shall

Fig. 9.4.

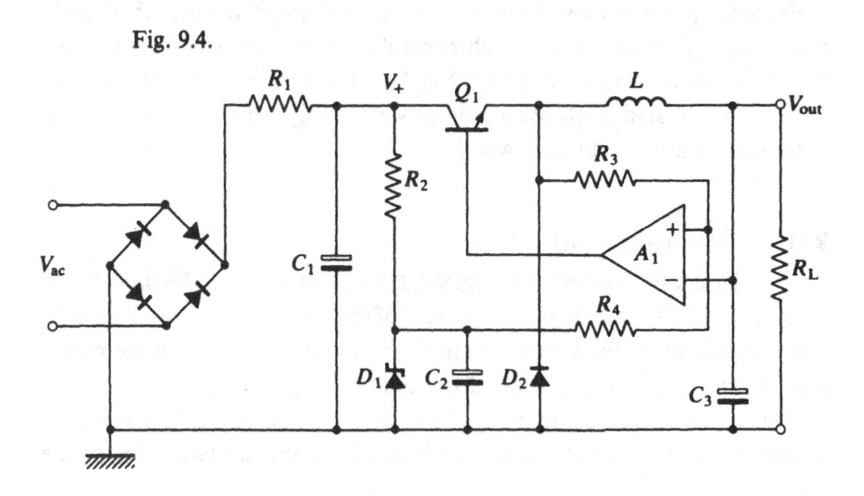

see, $R_3 \gg R_4$. It follows that the output of A_1 goes positive and turns on Q_1 so that the current in L begins to rise, initially, at a rate $di_L/dt = V_+/L$. The voltage across C_3 rises, as some of this current, i_L, flows into it, until V_{out} reaches the level

$$V_{out} = V_{ref} + (V_+ - V_{ref})R_4/(R_3 + R_4) \quad (9.1)$$

which can be made close to V_{ref} when $V_+ \gg V_{ref}$ and $R_3 \gg R_4$. Q_1 now turns off, because the voltage across the input terminals of A_1 has changed sign, and the current in L, now flowing through the diode D_2, begins to fall, approximately exponentially with time constant L/R_L. Note that the non-inverting input of A_1 is no longer at the voltage level given by the right-hand side of equation (9.1) but at the lower level, $V_{ref} - (V_{ref} - V_F)R_4/(R_3 + R_4)$, where V_F is the forward voltage of D_2. It follows that Q_1 will stay off until the output voltage has fallen to

$$V_{out} = V_{ref} - (V_{ref} - V_F)R_4/(R_3 + R_4) \quad (9.2)$$

whereupon Q_1 will be turned on again and the cycle repeated.

The circuit of Fig. 9.4 is thus a simple self-oscillating switched mode power supply which can operate very efficiently, even when V_+ is very much greater than V_{out}, because the power lost in the device which replaces the series regulator of the classical circuit, Fig. 9.2, is very small. The reason that this power loss is so small is that Q_1, in Fig. 9.4, is being used as a switch and the voltage across it is only $V_{CE(SAT)}$ when the pulse of current is passing. As the load on the power supply is increased, the rate at which V_{out} falls to the lower level, given by equation (9.2), increases. This means that the frequency of the pulses increases, and the width of the pulses increases also, because less of the pulsed current is available to charge C_3: more current is diverted into R_L.

From the point of view of Fig. 9.1, the control signal, in a switched mode power supply, is the signal which controls the rate and duty cycle of the switch: in the very simple circuit of Fig. 9.4, this is the output from A_1. In a more sophisticated switched mode power supply, this control signal is generated in a more definite way [2].

9.4 Changing sign

In the simple circuit of Fig. 9.4, the sign of V_{out} must be the same as the sign of V_+. By isolating the input side of the circuit from the output side, it is possible to invent a new circuit shape for the switched mode power supply which allows V_{out} to be of the opposite sign to V_+.

The circuit shape shown in Fig. 9.5 is a simple example of this new kind of circuit. Here C_1 provides d.c. isolation between the two sides of the circuit, and the drive to the base of Q_1 turns this transistor hard on for time τ, with period T. When Q_1 goes on, a voltage V_+ is developed across L_1 and the current in L_1 begins to rise at a rate V_+/L_1. When Q_1 goes off this current continues to flow into C_1, via D_1. Eventually, C_1 must become charged up to some constant voltage V_c, and we may write down the equations of the system in two pairs. The first pair apply when Q_1 is on,

$$V_+ = L_1 \mathrm{d}i_1/\mathrm{d}t \qquad (9.3)$$

$$V_c = L_2 \mathrm{d}i_2/\mathrm{d}t + i_2 R_L \qquad (9.4)$$

where the sign of V_c has been defined in Fig. 9.5, while the second pair,

$$V_+ - V_c = L_1 \mathrm{d}i_1/\mathrm{d}t \qquad (9.5)$$

$$0 = L_2 \mathrm{d}i_2/\mathrm{d}t + i_2 R_L \qquad (9.6)$$

apply when Q_1 is off. We have, of course, neglected the voltage drops across Q_1 and D_1 in comparison to other voltages in the circuit, and have assumed that C_1 is very large.

Equations (9.3) and (9.5) tell us that i_1 rises linearly, at a rate V_+/L_1, during the time τ, and then falls linearly, at a rate $(V_+ - V_c)/L_1$, for the rest of the cycle time $(T - \tau)$. It follows that the capacitor must charge up to a voltage, V_c, which is *greater* than V_+, otherwise $(V_+ - V_c)/L_1$ could not be negative and i_1 could not fluctuate about a mean level such that

$$(V_+/L_1)\tau + [(V_+ - V_c)/L_1](T - \tau) = 0 \qquad (9.7)$$

which tells us that

$$V_c = V_+ T/(T - \tau) \qquad (9.8)$$

We can find $\overline{V_{out}}$ by arguing that, for large values of L_1 and L_2, the fluctuations in i_1 and i_2 about their mean level are not very great and that $\overline{V_{out}} = -\overline{i_2} R_L$. Now for i_2 to fluctuate about a mean level, $\overline{i_2}$, we must have,

Fig. 9.5.

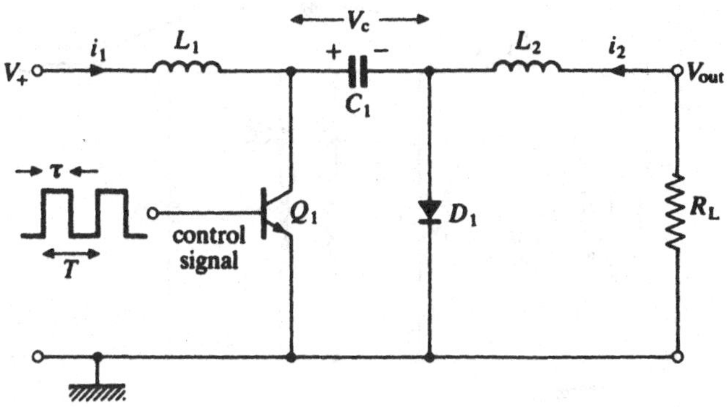

from equations (9.4) and (9.6),

$$[(V_c + \overline{V_{out}})/L_2]\tau + (\overline{V_{out}}/L_2)(T - \tau) = 0 \tag{9.9}$$

so that, substituting equation (9.8),

$$\overline{V_{out}} = -V_+\tau/(T-\tau) \tag{9.10}$$

Equation (9.10) tells us that V_{out} increases in magnitude as the width of the pulse, fed to the base of Q_1 in Fig. 9.5, is increased and $\overline{V_{out}} = -V_+$ when the input control signal becomes a square wave. This is the basis for the control signal in this circuit, as defined by Fig. 9.1. As the mark to space ratio of the input control signal is increased, $\overline{V_{out}}$ also increases.

9.5 An experimental circuit

The circuit shape shown in Fig. 9.5 was used by Ćuk to form the basis of a very interesting kind of switching regulator which can have a ripple free output [3]. Ćuk's circuit will be used for our experimental circuit on switched mode power supplies, because it involves some very important ideas which influence the entire field of power electronics [4].

The experimental circuit is shown in Fig. 9.6 and differs from the previous circuit in two ways. Firstly, an ideal transformer, T_1, has been placed in between the two sides of the circuit. This has a turns ratio of 2:1. Secondly, we allow L_1 and L_2 to become coupled, so that there is a mutual inductance, M, between them. C_1 and C_2 are assumed to be very large and equal to one another.

Let us examine the behaviour of this new circuit on a square wave control input which puts Q_1 on for time $\tau = T/2$. In the previous circuit, Fig. 9.5, we knew that this control input made V_c, equation (9.8), equal to

Fig. 9.6.

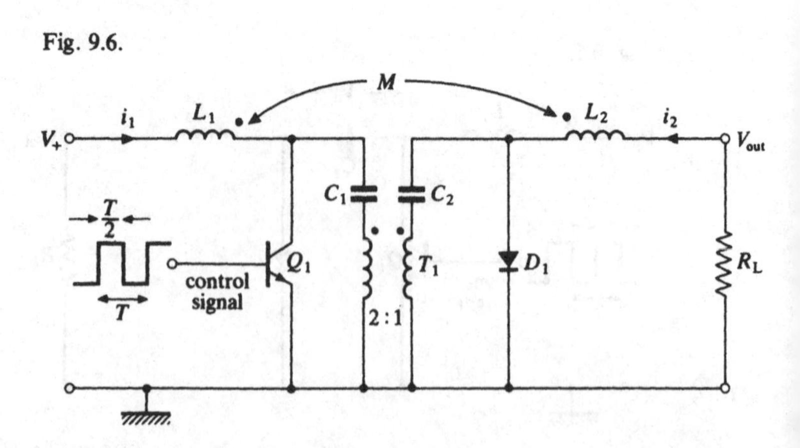

$2V_+$. In the new circuit, $C_1 = C_2$, and both capacitors charge up to a voltage $2V_+/3$, because of the transformer [5].

The equations of the circuit are now, when Q_1 is on,

$$V_+ = L_1 \mathrm{d}i_1/\mathrm{d}t + M\mathrm{d}i_2/\mathrm{d}t \quad (9.11)$$

$$V_+ = M\mathrm{d}i_1/\mathrm{d}t + L_2\mathrm{d}i_2/\mathrm{d}t + i_2 R_L \quad (9.12)$$

and

$$-V_+ = L_1 \mathrm{d}i_1/\mathrm{d}t + M\mathrm{d}i_2/\mathrm{d}t \quad (9.13)$$

$$0 = M\mathrm{d}i_1/\mathrm{d}t + L_2\mathrm{d}i_2/\mathrm{d}t + i'_2 R_L \quad (9.14)$$

when Q_1 is off.

Now these equations show that the circuit has the most interesting behaviour when we vary the value of M, which is very easily done experimentally if L_1 and L_2 are made simple air cored coils which can be placed end to end, or inside one another. If we view V_{out} on an oscilloscope, when the coupling between L_1 and L_2 is weak, a mean d.c. level is observed which has a triangular ripple voltage superimposed upon it at the same frequency as the square wave input signal.

If the coupling between L_1 and L_2 is now increased, by bringing the two coils closer together, always observing the correct sense, which is marked in Fig. 9.6 by means of the two dots, we observe that the amplitude of the ripple falls. Eventually, if we can make the coupling between L_1 and L_2 great enough, the ripple will vanish and then, on further increase in coupling, re-appear with the opposite phase, relative to the square wave drive signal.

This means that there must be a solution to equations (9.11) and (9.12), with (9.13) and (9.14), when we set $\mathrm{d}i_2/\mathrm{d}t = 0$ and solve for M in terms of $L_1 = L_2 = L$, that is the assumption of two equal inductors coupled together so that $M = k\sqrt{L_1 L_2} = kL$, where k is a coupling factor. It is a very simple matter to check that this is indeed true when $V_{out} = -V_+(M/L) = -V_+(1 - M/L)$, which can only be satisfied when $M/L = 1/2$, or $k = 0.5$ [6]. Details of the construction of Fig. 9.6 for experimental work are given in the Appendix. There are some interesting problems concerned with the design of the transformers which may be used in these circuits [7].

9.6 Audio amplifier output circuits

The audio amplifier is intended to provide a few watts of output, or in the case of some large scale applications up to perhaps 100 W, over a bandwidth from a few Hz to perhaps 100 kHz. It must do this using an input signal down at the microwatt level and introduce no distortion in the process. In this chapter we shall only deal with the output stages, or output circuit shapes, for audio amplifiers, because the low level circuits are very similar to the circuits already considered in Chapter 3. The output stage of an operational amplifier also involves similar circuit shapes to the ones considered here.

The introduction of power FETs [8] has made quite an impact on audio power amplifier design, and a recent review article [9] has dealt with this, and other particular problems of audio amplifier design, along with a good discussion of what is really needed for a detailed specification. Here, as in all parts of this book, the main concern is with the circuit shapes which turn up, and the reasons why these circuit shapes are used as the basis for a good circuit design.

Traditionally, the circuit shapes for audio amplifiers have been divided into three classes: A, AB and B. In class A, the current drawn from the d.c. power supply, see Fig. 9.1, and thus the power it delivers, is constant. This has considerable advantages from the point of view of low distortion, but it means that the amplifier must be very inefficient because the constant d.c. power input must always be much greater than the maximum power output. In audio applications the demand for maximum power output may be a very rare event.

In a class B amplifier we are dealing with a circuit shape of the kind already discussed in this book as Fig. 6.6. This circuit shape is repeated here as Fig. 9.7, this time using enhancement mode power MOST devices because this will illustrate the properties of this circuit more dramatically than the bipolar transistor version. Q_1 is an n-channel device and Q_2 a p-channel device.

Typical characteristic curves for Q_1 and Q_2 are sketched in Fig. 9.8, and these show at a glance that, in the connection of Fig. 9.7, it is impossible that both Q_1 and Q_2 pass current at the same time. V_{in} must exceed V_{out} by a

Fig. 9.7.

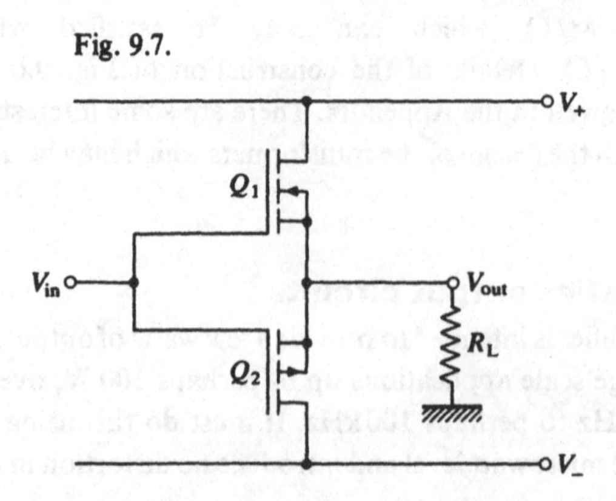

few volts in order to cause Q_1 to pass any current from V_+ into R_L. Similarly, V_{in} must be a few volts below V_{out} before Q_2 can come on to pull current out of R_L down to the negative rail, V_-. When $V_{in} = V_{out} = 0$, no current flows from either supply rail.

It follows that Fig. 9.7 is not a good basis to build upon for an amplifier which must work over a very wide range of signal level, and yet introduce negligible distortion. V_{in} must be of the order of a few volts before Fig. 9.7 produces any output at all. The other big problem with this circuit is that, in order to get V_{out} to approach the level of V_+ or V_-, it is necessary to take V_{in} *above* the level of V_+ or *below* the level of V_-.

9.7 The class AB output circuit

A better circuit shape than Fig. 9.7, for the output stage of an audio amplifier, is shown in Fig. 9.9. In this new circuit shape the gates of the two output transistors are no longer connected together. A voltage clamp, the circuit idea introduced in Fig. 6.13, consisting of Q_1 and the two resistors, R_1 and R_2, is introduced to bias both output transistors, Q_2 and Q_3, so that a small quiescent current flows from V_+ to V_- when $V_{out} = 0$. Fig. 9.8(*a*) shows that V_{in} will have to be slightly positive under this quiescent condition [11].

An output stage of the circuit shape shown in Fig. 9.9 is termed class AB because it lies in between class A, where the power supply currents are constant and independent of output level, and class B, where the power supply currents are both zero under conditions of zero output and increase linearly with output power.

Fig. 9.8. Typical characteristics of (*a*) an n-channel and (*b*) a p-channel enhancement mode power MOST. These are 30 V, $V_{DS(max)}$, devices which can dissipate 1 W continuously [10].

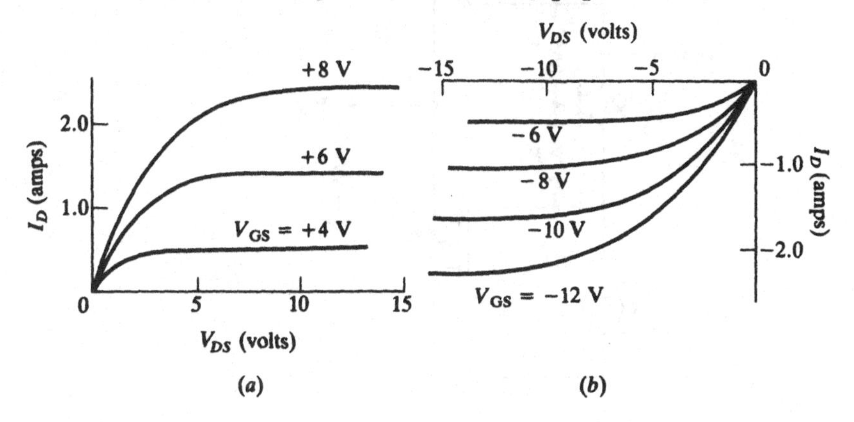

One of the main problems in the design of a class AB output stage is the definition of the quiescent current and its stability with changing temperature. If the output stage, for example in the reproduction of music, is suddenly called upon to deliver power at quite high level, the output devices, Q_2 and Q_3 in Fig. 9.9, may have to operate near maximum dissipation for a few minutes and will get hot. If the amplifier then has to return to the quiescent condition, or continue with the reproduction of a very quiet passage in the music, the bias setting, which was correct when Q_2 and Q_3 were cool, may now be quite incorrect. This problem is not so serious with MOS power transistors, because the temperature coefficient of gate to source voltage for constant drain current is fairly small. It is negative at low currents, that is destabilising, but changes sign at high currents. With bipolar devices the problem is far more serious because the temperature coefficient is large, $-2.1\ \mathrm{mV/^\circ C}$, at all current levels and, of course, destabilising in that currents increase as things get hot.

Methods of overcoming this problem of the temperature stability of the quiescent current need not change the circuit shape of Fig. 9.9. Resistors connected in between the output terminal and the sources of Q_2 and Q_3 will prevent any thermal instability of the circuit, but the inefficiency that these resistors introduce has to be removed by putting diodes across them. With bipolar output devices, the voltage clamp in Fig. 9.9, Q_1, R_1 and R_2, may be replaced by two simple silicon diodes, carrying a suitable forward current,

Fig. 9.9.

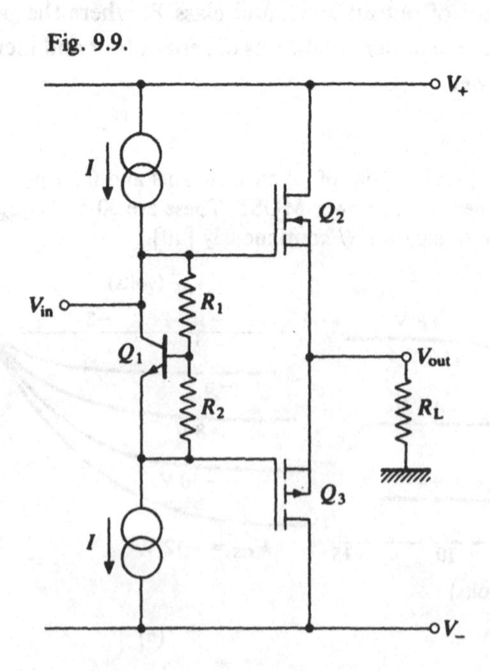

and then thermal stability can be arranged by very close thermal contact between these diodes and the output transistors.

There is still, however, a severe drawback to the output circuit shape of Fig. 9.9 and that is that the gates of Q_2 and Q_3 must be driven above V_+ and below V_- if the full potential output dynamic range is to be achieved. This problem is not so severe when Q_2 and Q_3 are bipolar devices, but it is still there. In order to overcome it, a radical change in circuit shape is needed, and this is what is proposed in the next section for our next experimental circuit.

9.8 An experimental audio amplifier

For an experimental circuit, let us look back at a new circuit shape which was introduced here as Fig. 6.14. This is shown again, with slight modifications, as Fig. 9.10.

In Fig. 9.10, two MOS power transistors are shown as the output devices of a power amplifier, supplying power to a load R_L. In this circuit, the two drains are connected together. When this circuit shape turned up in Chapter 6, as Fig. 6.14, the two gates were also joined together, and this is quite permissible with small MOS transistors and will define a quiescent current in Q_1 and Q_2 which depends upon the supply voltages and upon temperature. A very successful operational amplifier uses just such an output stage [12]. The quiescent current is easily found by a simultaneous solution of the non-linear characteristics, like those shown in Fig. 9.8. For MOS power transistors, however, the result would be too large a quiescent current and, in addition, this current would be too temperature dependent. Some more definite way of setting the quiescent current must be found.

The new circuit shape of Fig. 9.10 has, however, overcome the difficulty that remained with the previous circuit, Fig. 9.9. To take V_{out} close to V_-, in Fig. 9.10, V_{in2} must be taken towards zero while V_{in1} only needs to go close

Fig. 9.10.

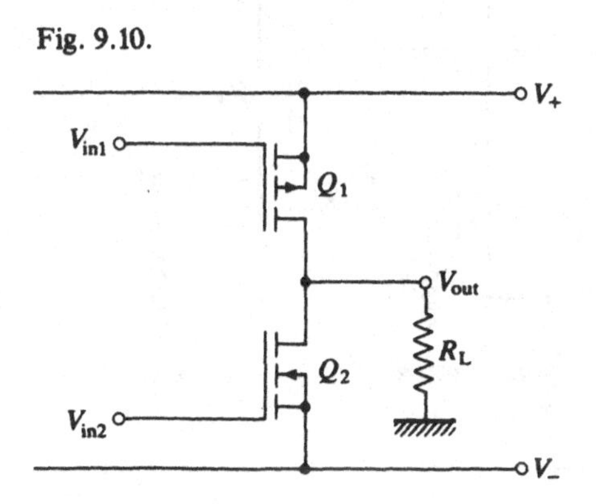

to V_+. Similarly, for V_{out} to approach V_+ it is now V_{in1} that must be taken down towards zero and V_{in2} that needs to approach V_-. All these changes in input level take place in between the levels of the power supply rails.

There is a further problem with Fig. 9.10, when we contrast it with Fig. 9.9, and that is its gain and output impedance are both high. Fig. 9.9 had 100% negative feedback from output to input which was intrinsic to its circuit shape: Fig. 9.9 is based upon the source follower circuit. This meant that the gain of the circuit shown in Fig. 9.9 was unity and that its output impedance was very low. We should like something like this from Fig. 9.10.

This means that if we wish to use the circuit shape of Fig. 9.10, with its advantage of high dynamic range at the output, we shall have to restore the feedback from output to input which has been lost in the change of circuit shape from Fig. 9.9 to Fig. 9.10. Fig. 9.11 shows a step in this direction where only the top half of the new circuit shape has been sketched. The lower half is left as a question mark for the moment. Fig. 9.11 thus acts as a stepping stone towards the final circuit we shall use.

In Fig. 9.11, the gate of the p-channel output device, Q_3, is driven by a simple common emitter npn device, Q_1. This is an obvious choice because it gives the required level shift from a V_{in}, which is at zero level, up to the positive level of the gate of Q_3. The base of Q_1 is a non-inverting input, however, so that negative feedback from the output of Q_3 must be applied to the emitter of Q_1. The most sensible way of doing this is, of course, to make Q_1 one half of a long tailed pair, and this is the solution shown in Fig. 9.11. V_{in} and V_{out} may now be both at zero level under conditions of zero input signal. The gain of the circuit shown in Fig. 9.11 will be $(R_4 + R_5)/R_4$.

Fig. 9.11.

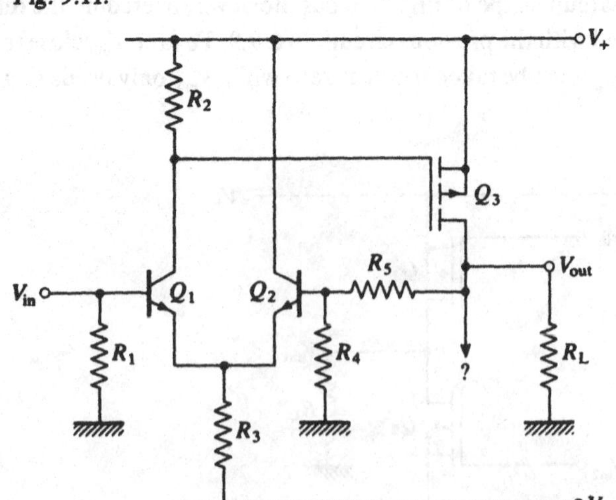

The output impedance of this new circuit shape may be found by removing R_L and asking what current, ΔI, must be drawn from the output terminal in order to pull the voltage level at this output terminal down by ΔV. As the gain of the long tailed pair is given by $I_{C1}R_2/2(kT/e)$, and $I_{C1}R_2$ is equal to the V_{GS} of Q_3, a simple approximation to $R_{out} = \Delta V/\Delta I$ is given by

$$R_{out} = 2(kT/e)(R_4 + R_5)/V_{GS\,3}g_{m\,3}R_4 \tag{9.15}$$

and this is only a fraction of an ohm in all practical cases. The load resistor, R_L, will be much greater than R_{out} [13].

The final problem to be solved is the setting of the quiescent current in the two output devices. Fig. 9.11 suggests a way of doing this when the base of Q_2 is considered as the virtual earth of a feedback amplifier. If a current I were drawn from this point, an identical current, I, would have to flow in R_5. Let us now add the lower half of the circuit and introduce this new idea.

Fig. 9.12 shows the final experimental circuit. Two small value resistors, R_{18} and R_{19}, are added to sense the quiescent current. The current, I, which was mentioned above, is provided by taking R_{13} down to the negative rail. This causes an identical current to flow in R_{16} which means, if

Fig. 9.12.

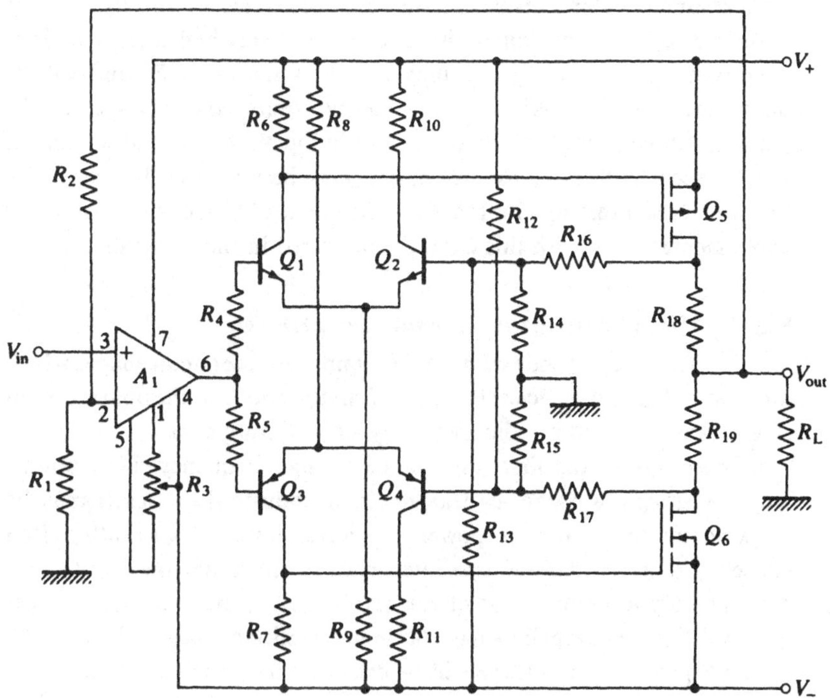

the entire circuit remains balanced with $V_{out} = 0$, that the quiescent current in Q_5 must be

$$I_Q = |V_-|R_{16}/R_{18}R_{13} \qquad (9.16)$$

The same quiescent current is arranged to flow on, through Q_6 to the negative rail, by making $R_{19} = R_{18}$, $R_{17} = R_{16}$ and $R_{12} = R_{13}$. The two supply voltages, V_+ and V_-, are, of course, equal in magnitude.

Note an interesting, and perhaps unexpected, addition to Fig. 9.11 when we move to the final circuit, Fig. 9.12. This is that Q_2, in Fig. 9.12, now has a collector load, R_{10}, which appears to be redundant. Similarly, Q_4 has a load, R_{11}. The reason for this is that discrete transistors are used in our circuit, and it is important to balance the power dissipation as much as possible. If this is not done, the offset voltages of the long tailed pairs change with output level, and the quiescent current is no longer well defined.

The voltage gain of the output stage is $(R_{16} + R_{14})/R_{14}$, which, of course, is made equal to $(R_{17} + R_{15})/R_{15}$. R_{12} and R_{13} are too high to affect the gain, while R_{18} and R_{19} are too low. If this output stage gain is made equal to about ten, the level of drive signal needed for the output stage is reasonably low and an overall voltage gain of about 100 may then be obtained by introducing a conventional operational amplifier, A_1 in Fig. 9.12, as an input stage. Overall feedback is applied *via* R_2 into R_1. The low frequency, or d.c., loop gain of this design is then very high indeed, and the condition $V_{out} = 0$, for $V_{in} = 0$, may be set by means of R_3 and will be maintained to very high precision, despite quite large changes in the working temperature of all six transistors in Fig. 9.12. Thermal stability of the quiescent current is also good, and may be improved by using well matched dual transistors for the long tailed pairs [14]. Constructional and experimental work with this circuit is discussed in the Appendix.

9.9 High frequency power amplifiers

This chapter ends with a brief summary of the circuit shapes which are used at high frequencies: the kind of circuits found in the final stages of radio transmitters and radio frequency heating equipment.

Efficiency is of the first importance in high frequency, high power, amplifier design; a broadcast transmitter or industrial r.f. heater may be working up to an output power of several hundred kilowatts. High efficiency must always be associated with a rather narrow bandwidth, perhaps only a few per cent of the carrier frequency, and such narrow bandwidth power amplifiers have become classified as class C, D, E, F, G, H and S [15]. Some of these will be discussed in the next section.

Narrow bandwidth is not a feature of all radio frequency power amplifiers, however, and it is not a feature which is called for in the lower power stages of a radio transmitter. These lower power stages, the stages which drive the final power amplifier, will usually be made wide-band so that changing transmitter frequency only calls for re-tuning of the final stage. This innovation belongs to the late fifties, and its importance at that time has been discussed by Baker [16].

Wide bandwidth power amplifiers for radio transmitters have become more important recently because certain systems of modulation call for very linear power amplifiers, and these combine well with wide bandwidth. Very frequent changes of operating frequency are also called for in many mobile and special low power communication systems, like radio telephones, and this again is far more straightforward if only the frequency of the first oscillator needs to be changed. The last two sections of this chapter take a brief look at the special circuit shapes found in wideband power amplifiers.

9.10 Narrow bandwidth, high efficiency, power amplifiers

The classical solution to the output power amplifier problem is termed 'class C'. In a class C amplifier, the active devices are effectively in series with a resonant load and a pulse of current is made to flow through these devices when the voltage across them is at a minimum. Fig. 9.13 shows a typical circuit shape for a class C amplifier using bipolar transistors.

The first thing to notice in Fig. 9.13 is that all the level shifting problems, which were so much to the fore in our previous circuits, can be removed in narrow band amplifiers by means of transformers. If Fig. 9.13 represents the output stage of a small VHF transmitter, L_1 and L_2, and L_3 and L_4, would be of a few turns, interwound on a simple cylindrical former.

The class C amplifier shown in Fig. 9.13 works with the collector voltages of Q_1 and Q_2 rising very close to 2 V_+. The input drive must be taken up to a level which makes this happen, and then any amplitude modulation of this output stage must be done by varying V_+. Q_1 is then drawing a pulse of current when Q_2 is off, and *vice versa*. When the pulse of current flows in either Q_1 or Q_2 the V_{CE} of the device is only a volt or so and power dissipation is very low. Efficiencies of well over 90% can be obtained with these circuits, provided the operating frequency is low enough to allow the turn-on and turn-off times of Q_1 and Q_2 to be really negligible.

The mean base current of Q_1 and Q_2, in Fig. 9.13, flows up R_1 and R_2 to cause a mean negative bias voltage to appear across the emitter junctions of both devices. Capacitors C_1 and C_2, and C_3 and C_4, perform a simple impedance transformation between the drive input and the input impedance of the transistors. This input circuit would be of fairly wide bandwidth. C_5 and C_6 perform a similar transformation between the output impedance and the load.

An alternative to class C has become known as class D. High efficiency is obtained in class D by using the active devices as switches, and this can be done in two distinct ways: voltage switching or current switching.

A voltage switching class D amplifier could have a circuit shape virtually identical to Fig. 9.10. The two inputs, $V_{in\,1}$ and $V_{in\,2}$, would be transformer coupled and arranged to switch Q_1 on for half the time, when Q_2 was off, and *vice versa*. This would produce a square wave across R_L, in Fig. 9.10, but in the radio frequency class D power amplifier, the load is no longer a simple resistor but a series resonant circuit: L, C and R_L. This is tuned to the input drive frequency. The current in R_L is then sinusoidal and has a peak amplitude $(4/\pi)V_+/R_L$. Efficiencies of well over 90% are easily obtained with these amplifiers [17].

The current switching class D amplifier is particularly interesting in that it has a very early origin [18] and may be one of the earliest electronic circuits to get away from the constraint of using a constant voltage power supply, and use a constant current power supply instead. The circuit shape is shown in Fig. 9.14.

Fig. 9.13.

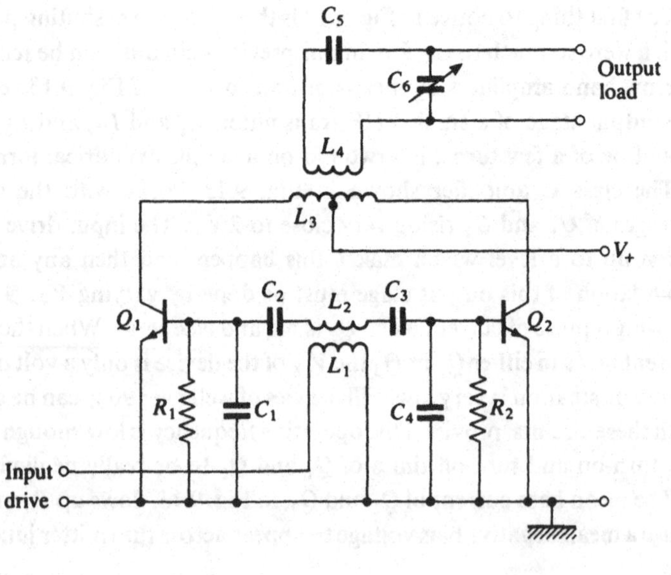

At first glance, Fig. 9.14 looks very similar to the class C amplifier shown in Fig. 9.13. However, in the class D case, Q_1 and Q_2 are not pulsed on for only a small fraction of the cycle, but switched on for the whole half cycle. This means that it is quite impossible to supply the centre tap of the inductor, L, with a constant voltage. The constant current supply, I_{dc}, allows the voltage at this centre tap to rise periodically to $\hat{V}_L/2$, following a half sine wave each half period. It is well worthwhile sketching the waveforms for this deceptively simple circuit, particularly the waveform of the current which flows in the capacitor.

Finally, among the high efficiency, narrow band, power amplifiers, class E and F should be mentioned. The origins of these circuits have been discussed by Tyler [19]. The idea is to make it possible for the voltage across the active devices to be a square wave, while the voltage across the resonant load is, of course, sinusoidal, by inserting filters in between the switching devices and the load. The high efficiency of switching is then obtained in a more defined way than with simple class D because the switching speed of the devices used can be enhanced by correct design of the filter networks [20].

9.11 Wide bandwidth, high frequency, power amplifiers

With the wide-band power amplifiers, discussed in Section 9.9, we may find circuit shapes very similar to the class AB circuits described in Sections 9.7 and 9.8. With these circuits, efficiencies up to about 70% are possible at low frequencies. For example, Meier [21] describes an amplifier with a bandwidth from 1.6 MHz to 30 MHz which can produce 300 W output. This design uses n-channel enhancement mode power MOS transistors. At high frequencies it is very difficult to make a wide-band or

Fig. 9.14.

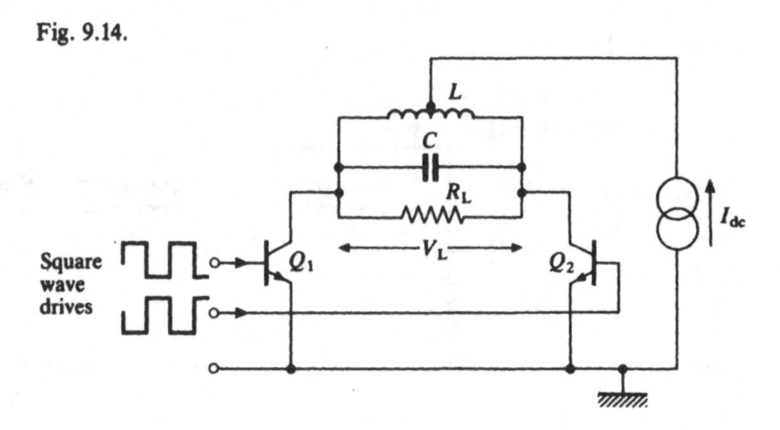

linear power amplifier with an efficiency better than perhaps 50%, but when advanced modulation methods make VHF and UHF communications possible with transmitter powers of only a few watts, this is no great problem.

The class AB amplifiers used at audio frequencies, for example Figs. 9.9, 9.10 and 9.12, use complementary devices in their output stages. In high frequency wide-band amplifiers of the class AB, or B, type we are more likely to find devices of the same polarity being used, usually npn bipolar or n-channel enhancement power devices, because these particular polarities make the higher power, voltage, current and frequency performance possible.

Circuit shapes for class AB amplifiers which do use pairs of identical devices in the output stage are quite interesting. An example is shown in Fig. 9.15, where Q_1 and Q_2 would both be passing a quiescent current of about 10% of $I_{C(max)}$. This would be arranged by means of a separate biassing circuit, not shown in Fig. 9.15, which would involve some kind of temperature sensor, such as a diode, as an integral part of the power transistor [22]. The input drive, in Fig. 9.15, would then be arranged to drive Q_1 on for the first half cycle and Q_2 on for the second half cycle, continuously, always ensuring that the V_{CE} of both devices falls to only about 1 volt, at minimum, and then rises during the time the transistor is off to a maximum of near $2V_+$. It is essential that all amplifiers of this linear kind run at full power all the time, otherwise the dissipation in Q_1 and Q_2 may be too high. These amplifiers handle an input which is either frequency modulated or single sideband: modulation systems which involve constant radiated power from the transmitter.

Fig. 9.15.

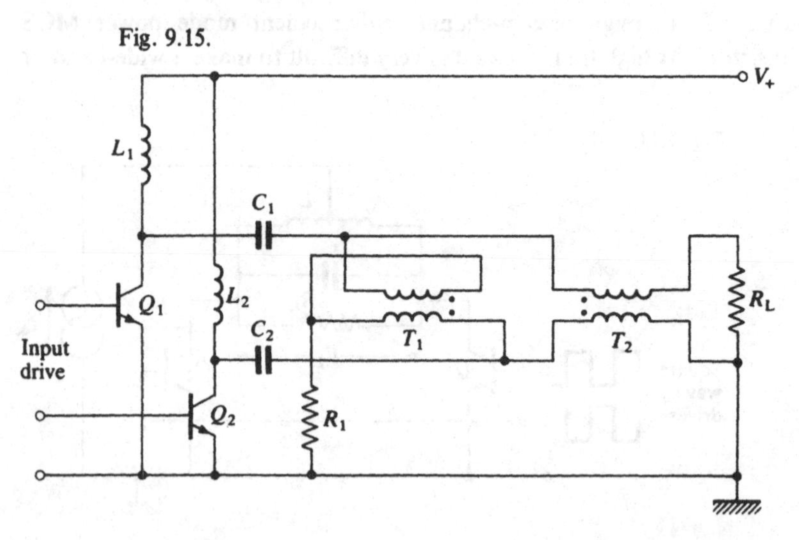

L_1 and L_2, in Fig. 9.15, have a reactance much greater than the resistive load seen by Q_1 and Q_2, which, in this particular example, is $R_L/2$. Naturally, the resistance of L_1 and L_2 must be very much smaller than $R_L/2$. This specification for L_1 and L_2 must be maintained over the entire bandwidth of the amplifier.

After simple d.c. isolation, *via* C_1 and C_2, the outputs from Q_1 and Q_2 arrive at T_1, which combines these two outputs and gives an impedance transformation of 1:2. In English, the biological term 'hybrid' has become accepted to describe power combiners of this kind. The action of T_1 may be seen by considering the pulsating currents flowing alternately in Q_1 and Q_2. When Q_1 draws I_{C1} we will find $I_{C1}/2$ being drawn from the top of R_L, *via* the balanced to unbalanced, 1:1, transformer, T_2. The other half of I_{C1} is drawn from T_1, so that $I_{C1}/2$ flows into the top dotted end of T_1. It follows that $I_{C1}/2$ must then flow out of the lower dotted end of T_1, in Fig. 9.15, and thus supply the bottom of R_L, the current in Q_2 being zero during this half cycle. The full current, I_{C1}, flows up R_1, and for Q_1 to see the correct load we must make $R_1 = R_L/4$.

On the other half cycle, when Q_2 conducts and Q_1 is off, the argument is reversed but the direction of the current in R_1 is still the same. However, R_1 cannot have a d.c. level across it and, in fact, carries the sum of the alternating components of I_{C1} and I_{C2}. Thus the fundamental component vanishes and only a predominantly second harmonic current flows in R_1. Even under pure class B conditions, the power dissipation in R_1 is less than one eighth of the output power into R_L. As a class B amplifier can be operating near 75% efficiency, this means that wide-band performance with the circuit shape of Fig. 9.15 can be obtained with an efficiency over 50%. Some practical designs of wide-band power amplifiers using these techniques have been discussed by Oxner [23] and by Krauss *et al.* [24].

9.12 The distributed amplifier

This chapter closes with a circuit shape which has turned up repeatedly in the literature since it was first described by Percival [25] more than fifty years ago: the distributed amplifier. The circuit shape is shown in Fig. 9.16 and involves two transmission lines. The lower line in Fig. 9.16 accepts the input signal and power is transferred, after amplification, from this input line to the output line, at the top of Fig. 9.16, by a number of amplifiers: Q_1, Q_2; Q_3, Q_4; Q_5, Q_6; Q_7, Q_8; and Q_9, Q_{10}. For our example, we have chosen an amplifier circuit which was considered in Chapter 2: Fig. 2.3. Any amplifying circuit with negligible feedback from output to input would be suitable, however.

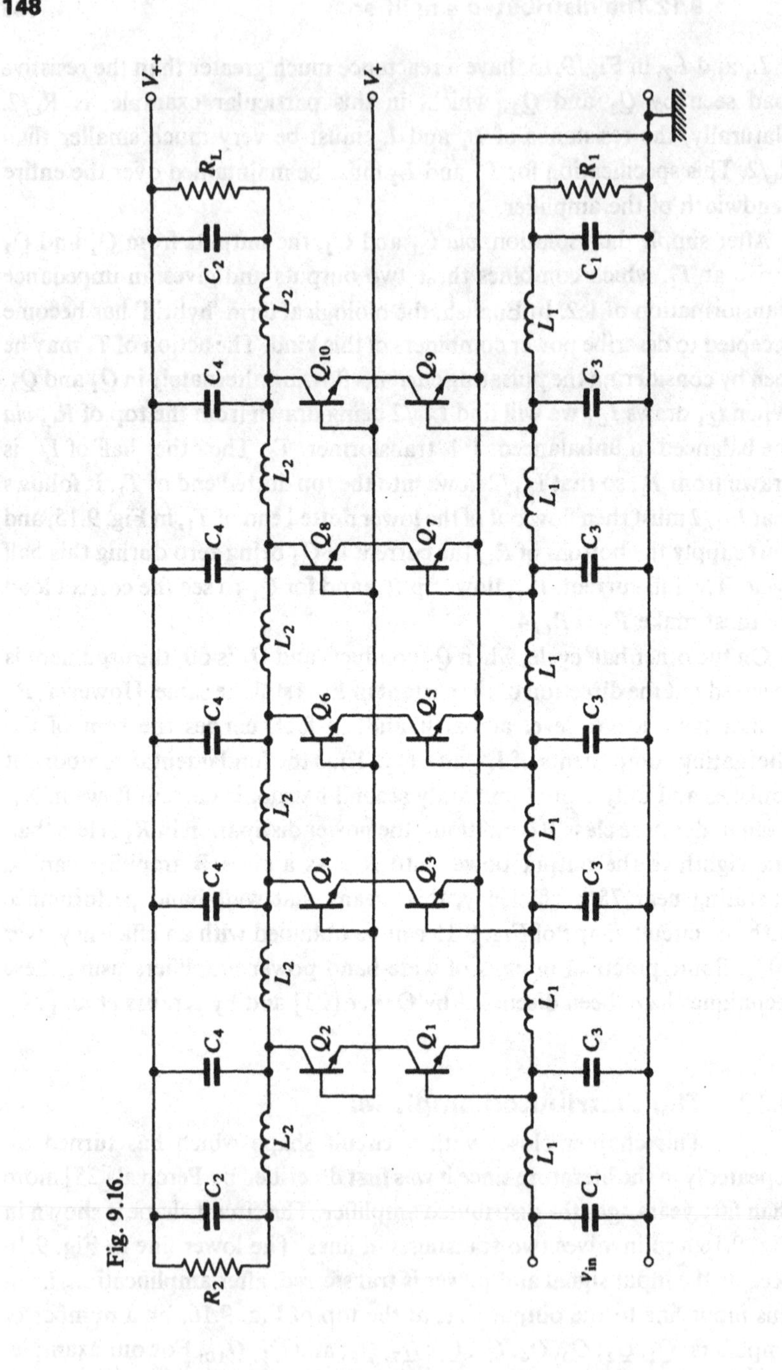

Fig. 9.16.

In Fig. 9.16, the two transmission lines exhibit the same phase velocity. If the input capacitances of Q_1, Q_3, Q_5, Q_7 and Q_9 could be neglected, we would have $C_1 = C_3/2$. However, one of the main features of the distributed amplifier is that the input capacitance of each stage becomes part of the lumped capacitance making up the lower line, so that each C_3 is reduced by an amount C_{in}. Similarly, the output capacitance of each stage is in parallel with each C_4. R_1 terminates the input line, which receives the input signal from a source impedance equal to R_1. The output line is terminated at both ends: R_2 on the left and the output load, R_L, on the right. Fig. 9.16 has been simplified as much as possible by omitting most of the biassing arrangements for the transistors.

The point about distributed amplifier technique is that it is an attempt to overcome the intrinsic gain–bandwidth limitations of the amplifying circuits we have considered up to now. That is why the distributed amplifier belongs in our chapter on power amplifiers because it is with power devices that this intrinsic limitation may still be important. In earlier times, distributed amplifiers were used for some small signal applications which needed wide bandwidth.

What, then, is the intrinsic limitation on gain–bandwidth product in a simple amplifier? An estimate of this may be made by using the simplest possible model for the amplifying device: an input capacitance, C_{in}, and an output current generator, $g_m v_{in}$, in parallel with a capacitance, C_{out}. Connected in series with a load resistor, R_L, and driving a second identical device, such an amplifying device will give a voltage gain $G_0 = g_m R_L$ at low frequency. This gain will drop by 3 db when the input frequency reaches

$$f_0 = 1/2\pi(C_{in} + C_{out})R_L \tag{9.17}$$

and it follows that the gain–bandwidth product is given by

$$G_0 f_0 = g_m/2\pi(C_{in} + C_{out}) \tag{9.18}$$

which depends upon the device only, not upon the circuit shape.

A distributed amplifier circuit, like the one shown in Fig. 9.16, attempts to obtain a gain–bandwidth product well in excess of that given by equation (9.18), by causing the input power to pass from device to device, as it travels down the input line, before it is finally dissipated in the termination, R_1. The fact that this final dissipation of the input power in a resistor must take place in all wide-band amplifiers was discussed at some length in Section 5.6.

These fundamental aspects of the distributed amplifier approach to wide-band power amplifier design have been discussed by Sosin in an important paper [26]. Sosin deals with the interesting case in which the characteristic impedance of the output line is reduced as we move towards the output termination. This can improve the efficiency of the amplifier considerably. The basic circuit shape shown here as Fig. 9.16 would, of course, be a class A amplifier with very poor efficiency. Push–pull distributed amplifier designs are a simple extension of the same idea. The circuit shape is particularly suitable for MOS power devices [27].

Notes

1 A good review article on switched mode power supplies has been given by R.S. Olla, *Electronics*, **46**, No. 17, 91–5, 16 August 1973. Written at the time when the switched mode power supply was just becoming popular, this article puts the development in a good historical setting.

2 An excellent book which deals with switched mode power supplies is *Design of Solid State Power Supplies*, by E.R. Hnatek, Van Nostrand Rheinhold Co., New York, 2nd edn, 1981.

3 S. Ćuk, *IEEE Trans. Magnetics*, **MAG–19**, 57–75 and 75–83, 1983.

4 S. Ćuk and R.D. Middlebrook, *IEEE Trans. Ind. Electr.* **IE–30**, 10–19 and 19–29, 1983. The first of these two papers includes a description of a 40 W audio amplifier using switched mode technique.

5 To see why this is, consider the case when D_1 is on and Q_1 is off. If $C_1 = C_2 = C$, then C_2 appears as $C/2$ across the primary of T_1 and while C_1 will have 1/3 of $2V_+$ across it, the primary will have the other 2/3. On the secondary side, this voltage is put back to 1/3.

6 This is, of course, a particular solution for the special case of a 2:1 transformer ratio. A full course on these important developments has been published in three volumes: *Advances in Switched-Mode Power Conversion*, edited by R.D. Middlebrook and S. Ćuk, TESLAco, Pasadena, 1983.

7 D.V. Otto and P. Glucina, *Electr. Lett.*, **22**, 20–1, 1986.

8 E.S. Oxner, *Power FETs and Their Applications*, Prentice-Hall Inc., Englewood Cliffs, 1982.

9 J.L. Lindsay-Hood, *Wireless World*, **88**, No. 1558, 63–6, and **88**, No. 1559, 28–32, 1982.

10 The January 1985 *MOSPOWER DATABOOK*, Siliconix Inc., Santa Clara, is a valuable source of characteristics for a wide range of MOS power transistors. The curves shown in Fig. 9.8 correspond approximately to the VNO300M and VPO300M.

11 The circuit shape of Fig. 9.9 is used in the output stage of a design published by E. Borbely, *Wireless World*, **89**, No. 1566, 69–75, March 1983.

12 This is the RCA CA3130. RCA Data File No. 817.

13 In our experimental circuit, R_L is chosen to suit the devices. In a real design problem, the devices are chosen to suit the specified value of R_L and the maximum power that must be delivered to R_L.

14 This method of setting the quiescent current has been described, for a bipolar class AB output stage, using two operational amplifiers instead of simple long tailed pairs: T.H. O'Dell, *Electronic Engineering*, **57**, No. 706, 39, October 1985.

15 There is a table on p. 472 of *Solid State Radio Engineering*, by H.L. Krauss, C.W. Bostian and F.H. Raab, John Wiley, New York, 1980, which explains all the classifications which have been adopted in this field.
16 W.J. Baker, *A History of the Marconi Company*, Methuen, 1970, pp. 378–9 and 392–3.
17 Instead of two complementary devices, as shown in Fig. 9.10, a class D r.f. amplifier would be more likely to use two identical devices, as the ease of transformer coupling the input drive signal makes this straightforward. A further step along these lines is then to use four devices for voltage switching, connected as a bridge. A good review of these developments has been given by H. Ikeda, *IEEE Trans. on Broadcasting*, **BC–26**, No. 4, 99–112, 1980. Ikeda correctly predicted that MOSFET power devices would take solid state broadcast transmitters up to the 50 kW output level by 1986.
18 D.C. Prince and F.B. Vogdes, *Proc. IRE*, **12**, 623–50, 1924.
19 V.J. Tyler, *Marconi Review*, **21**, No. 130, 96–109, 1958.
20 N.O. Sokal and A.D. Sokal, *IEEE J. Sol. State Circ.*, **SC–10**, 168–76, 1975.
21 B. Meier, *RF Design*, November–December 1984, pp. 21–4.
22 For example, see J. Ling, *Electronic Components and Applications*, **3**, No. 4, 210–33, 1981. This article describes a 400 W class AB wide-band amplifier, 1.6 MHz to 30 MHz, in which the output bipolar transistors and their bias circuit are all mounted on the same heat sink. All constructional details are given with excellent colour photographs of the hardware.
23 Note 8 above, Fig. 10–10.
24 Note 15 above, Fig. 12–15c.
25 W.S. Percival, British Pat. No. 460562, 24 July 1935.
26 B.M. Sosin, *Proc. IEE*, **110**, 1374–84, 1963.
27 Note 8 above, Fig. 10–6.

10

Theory and practice

10.1 Introduction

After the introductory chapter of this book, on the problem of what is meant by design in general, eight chapters have been given dealing with particular areas of electronic circuit design, outlining a laboratory-based course.

These eight chapters have only dealt with a fraction of the field. A further eight, on phase locked loops, on switched capacitor circuits, on mixers and demodulators, on memory circuits, on waveform generators, on low noise amplifiers, on digital to analog interfaces and on very low level d.c. amplifiers, would enable us to look at even more circuit shapes and ideas, but the field would still be far from covered.

It is now time, however, to look back at what has been done and to ask if anything useful has been achieved. Throughout this book, we have concentrated upon the idea of circuit shape. As this was first put forward in Chapter 1, circuit shape was considered to be the first essential step in the design process: a preliminary sketch of our idea on how something should be put together. Does this mean that the *practical* side of electronic circuit design has been emphasised at the expense of the *theoretical*? What, exactly, do these two frequently used terms really mean?

10.2 Theory

If we look back at the last circuit that was considered, Fig. 9.16, it seems beyond question that the people who first came up with this new circuit shape had a considerable theoretical background in what is usually called *transmission line theory*. This is a classical branch of electrical engineering theory; it is not very much emphasised in engineering education at the time of writing, so that if some person were to re-invent the distributed amplifier today it would be very probable that, for some reason, classical transmission line theory had caught that person's interest.

Why does Fig. 9.16 suggest that this theoretical background must have been essential? Looking at the circuit, and remembering that the whole point of this circuit shape is to obtain very wide bandwidth, we note that the propagation delay from v_{in} to R_L is the same for all possible amplification paths through the circuit. If we trace the signal from v_{in} to R_L via Q_1 and Q_2, or Q_3 and Q_4, and so on, the propagation delay is always the same because both the lower and the upper transmission lines are designed to have the same phase velocity. This feature is an essential condition of a wide-band, distributed, system, but knowledge at that level does not belong to intuition. Compare the behaviour that would be expected if the output of the distributed amplifier, shown in Fig. 9.16, were taken from R_2, instead of R_L: the propagation delay *via* each cascode amplifier is now completely different.

Another feature of Fig. 9.16, which would be tacitly obvious to an engineer who was very familiar with *lumped* transmission line theory, is that there must be a high frequency cut-off in this amplifier determined by the product L_1C_1, which is, of course, equal to L_2C_2. A *uniform* transmission line, provided it is a TEM structure, like microstrip or coaxial line, does not, to a first approximation, have a cut-off frequency until the wavelength approaches the smallest dimension of the structure, such as its diameter. However, the whole point of the original invention, and the use of a lumped transmission line, was that the capacitors, in Fig. 9.16, could be made up, partially, by means of the input and output capacitances of the amplifying devices.

The same argument could be put forward for all the circuit shapes which have been presented in the previous chapters: a considerable theoretical background is needed before the idea would appear. For example, the power amplifier of Fig. 9.12 depends upon feedback amplifier theory very considerably, as did all the circuits discussed in Chapter 3. The circuits of Chapter 8 involved semiconductor device theory, so did the circuits of Chapters 4, 5 and 6. In Chapters 2 and 7 we used classical *LCR* circuit analysis.

Books dealing with electronic circuits usually deal with this theoretical background far more than with, what might be called, practical circuits. It is this theory, after all, which forms the basis of student examinations. The actual level at which this theory must be either conscious, or actually understood, before new electronic circuit shapes can be thought out is a question which can only be asked here. This is obviously an area of research which would be very difficult but very interesting.

10.3 Science and technology

The interaction between theory and practice, touched upon briefly in the previous section, brings us to another relationship, and one with which it is often confused: the interaction between science and technology. First of all, what is meant here by these two terms?

Science is a human activity which is concerned with understanding nature. For example, finding out why some metals, like aluminium or chromium, have the property that their oxides form a very tenacious protective film on the surface of these metals. Other metals, like iron, do not appear to have this property.

Technology is concerned with the human activity of manufacture [1]. Silicon, for example, also shows the remarkable property, commented on above for aluminium and chromium, that its oxide adheres very strongly to its surface. It is not essential to have a deep scientific understanding of exactly why silicon has this property in order to use it as the basis of silicon integrated circuit photolithographic technology. We know that germanium does not have an oxide with this property, and this is one reason why we do not have germanium integrated circuits.

Nevertheless, a person working for a semiconductor manufacturer who attempts to get a deeper understanding of why silicon oxide does adhere so well to silicon is most definitely behaving as a scientist. That same person, who moves on to the problem of finding good ways of etching through this oxide in order to make integrated circuits, becomes a technologist.

This brings us back to Chapter 1, and the point made in Section 1.3 that the idea of technology being some kind of 'applied science' is incorrect. There are some very interesting facts which should be considered in connection with the distinction between science and technology, and these facts have been put forward very clearly by Price [2], who has made a detailed study of the way people behave who are concerned with science on the one hand, and technology on the other. The behaviour which is most easy to observe and measure is the way these people go about the business of publication.

Price found that scientists take their own publications very seriously, particularly how many they have. The information in these publications, however, is already disseminated by other means: correspondence, meetings, discussions and so on. In contrast, only a tiny minority of technologists publish at all. Technologists certainly take their patents very seriously. They are also very concerned about how many they have, but they certainly do not want everybody to read these patents straight away. Price sums this up well when he writes 'To put it in a nutshell, albeit in exaggerated form, *the scientist wants to write but not read, and the technologist wants to read but not write*'. However, Price points out that an individual may well alternate between the two camps all the time.

Why is there this difference in behaviour between scientists and technologists? The explanation must be that the scientists see their publications as the record of their work, and as some measure of its worth. The technologists see their artifacts and processes, the actual hardware, as marking their contribution to whatever society they feel at home in. An individual technologist may often be able to write a scientific paper about some aspect of the work that was involved in the production process, and this will be accepted as a mark that this individual has some status among both scientists and technologists. Many names would cross the mind of an electronic circuits specialist at this point.

That technologists read a great amount is not surprising. They are looking for ideas on how things are done. They will read scientific papers, but usually these are scientific papers of the kind referred to above: written by technologists and inspired by a technical problem. They will read the rare technical publications of their fellow technologists, but what exactly are these technical publications about? That is the topic of the next section, but before considering this there is one final aspect of the relationship between science and technology that should be mentioned.

One very big input into science from technology is instrumentation. Price again puts this well: 'Here we have a technology which has been crucial to the advancement of science and perhaps a domain of strong interaction much more important than the apparently weak and infrequent "application" of science'. To an electronic circuits specialist, this seems a very sensible statement. Technologists were attempting to develop computers long before the scientific community found life to be impossible without them. High speed electronic circuits, so essential for the nuclear physicist, were a main stream activity in television development groups in the thirties. The awe inspiring radio telescopes of our time, with their gigantic size, extremely low noise figure circuits and sophisticated control systems, came on the scene with little hint of the major scientific discoveries that would be made with them. As Price concludes: 'If, for example, it is the social needs that produce the new technology and the technology which in turn gives rise to a new scientific understanding, we shall have a basis which is quite the opposite of that naively understood in the political arena'.

10.4 Technique

Let us now return to the question which was left unanswered above, and find out what it is that technologists do publish when they are not behaving as scientists and publishing scientific papers.

The answer that will be given here is to give the name 'technique' to this published material. By technique we mean an account of how something is produced, but this kind of technical information is found on a number of levels. Highly detailed accounts of technique, publications which give every detail of how an artifact was manufactured or how a process was set up, are certainly not made generally available, except, in some rare cases, after a very long time and when these details have only historical interest. This highest level of technique will be called 'practice' here, and is discussed briefly in the last section of this chapter.

At the simplest level of technique we find books such as this one, and nearly all the references that have been given here. These form the sketch books or pattern books of technical ideas, giving sufficient detail to allow the reader to make an artifact which will be similar to the one the original author described.

At the next level of published technique, we find those rare publications which describe the design and construction of something in detail. There have been some examples here, in previous chapters. In Chapter 5, note 1, the book edited by Sandbank is an example. In Chapter 6, note 4 listed a remarkable series of IBM papers which dealt with manufacturing technique as well as circuit detail, but these IBM papers were examples of technical publication many years after the work was started: a very large number of person-years needs to be invested to get a set-up like that. In Chapter 7, the paper by Baxandall (note 13) is a good example of giving detail, and was used as the basis of the second experimental circuit for that chapter. Finally, we must mention the paper by Ling (Chapter 9, note 22) as an outstanding example of detailed published electronic technique.

10.5 Practice

The very detailed technical publications, discussed in the previous section, are not only rare but have another special feature. This is that these papers either appear to be full of detail, like the IBM papers which were mentioned above, because an unusually wide field of design and manufacturing technique has been covered, or the detail is really complete only because a small circuit has been described and this circuit is also constructed in such a way that an individual author, or perhaps a few authors, could completely cover the work involved. The experimental circuits described in this book are of the latter kind.

However, when we look at what will be termed practice here, there is still a considerable amount of missing information. A circuit which has been built following exactly the same diagram, layout and constructional technique given in a publication will still show individuality: a second circuit built in apparently just the same way will not behave exactly as the first one did. The reasons for this are, of course, well known and understood. The range of behaviour may be predicted with considerable accuracy, provided an accurate model for simulation can be put forward. Pederson's review [3] of circuit simulation is a valuable guide to the literature in this connection, but such detail is not the prime concern at the level we have been considering in this book. Understanding electronic circuit design, to begin with, brings up circuit shape as the first concern and then input–output behaviour as the main way of seeing what has been achieved.

Finally, electronic circuits are used to build systems, and it is at the system level that very small shifts in circuit behaviour, shifts away from the intended specification, may introduce very serious problems. Computing systems, where very large numbers of individual, and identical, circuits are combined, may show this kind of difficulty, and a radical change in the basic circuit shape will have to be made to remove a problem which was not noticed, or anticipated, when the designer was working at the simple circuit level [4]. It is this kind of situation that illustrates the real distinction between practice and technique. But design involves a continuous circulation of thought and action: technique to practice to theory to technique. If all goes well, the circulation should not come back to the same point, but lead to a better way of doing things.

Notes

1 In contemporary society, technology is concerned not only with manufacturing and processing, but also with services and administration. This has been discussed in depth by Jacques Ellul in his books *The Technological Society*, Jonathan Cape, 1965, and *The Technological System*, Continuum Publ. Corp., New York, 1980, and also by Simon Nora and Alain Minc in their *The Computerisation of Society*, MIT Press, Cambridge, Mass., 1980.

2 The quotations which follow are from D.J. Price, *Tech. Cult.*, **6**, 562, 1965, and **15**, 47–8, 1974.

3 D. O. Pederson, *IEEE Trans. Circ. Syst.*, **CAS–31**, 103–11, 1984.

4 An excellent historical review, 'Design automation in IBM', has been published in the *IBM J. Res. Dev.*, **25**, 631–46, 1981. This gives some examples of how these problems with large computing systems can be dealt with. For a review of current work, see the entire No. 5 issue of Volume 28 of the same journal.

Appendix

This Appendix gives the details needed to construct the thirteen experimental circuits which have been described in the book, and gives some further suggestions for measurements.

If an introduction to constructional technique itself is needed, the reader should refer to chapter 12 of *The Art of Electronics*, by P. Horowitz and W. Hill, Cambridge University Press, 1980. This gives an excellent account of the available methods, and is well illustrated with photographs.

Fig. 1.2 This is a very interesting circuit to experiment with, particularly if you have a really wide-bandwidth oscilloscope: at least 250 MHz. The circuit then shows you the dramatic difference between general purpose transistors and really high frequency devices. The circuit should be made as compact as possible and it is essential to decouple the power supply lines with two capacitors, not shown in Fig. 1.2, from V_+ to ground and V_- to ground. These should be about 0.33 μF and, again, mounted as close as possible to the rest of the circuit on the circuit board. Try different values, and kinds, of decoupling capacitor and try leaving them off altogether.

Use a $\pm$15 V supply. If a floating, centre tapped, supply is available it is a good idea to make V_+ the true ground line and then make $R_4 = 50\,\Omega$ so that 50 Ω coaxial cable can be used to take the output directly to the wide-band oscilloscope. Otherwise, use a high impedance probe.

Other sensible values, giving an oscillation frequency of a few MHz, are $R_1 = R_3 = 220\,\Omega$, $R_2 = 1.2$ k, $R_5 = 1.8$ k and $C_1 = 2200$ pF. Compare the waveforms using two fairly fast devices (for example, 2N2369A) with the waveforms obtained using slower, higher current, devices (for example, 2N2222A). Check that the transistors do not saturate.

Fig. 2.4 The component values for this circuit are given in the text and the transistor can be any dual gate protected device, for example the 3N211 or BFR84. The two coils should be mounted with their axes normal to the plane of the circuit board, about 40 mm apart with the transistor in between. Make the top of each coil the r.f. grounded end.

The circuit will not work in the way described in the text unless a thin metal screen is placed between the two coils, preferably right across the device, and grounded. Make this about 60 mm wide and 40 mm high, but experiment with and without the screen in position.

Fig. 2.10 In this circuit L_1 and L_2 are both single layer, 40 T, 150 μm diameter wire, on 6 mm formers, 6 mm long. The formers are again with adjustable slugs and should be mounted, again with a screen in between, as described for Fig. 2.4. This is for operation close to 10 MHz.

The other circuit values are $C_1 = 33$ pF, $C_2 = 25$ pF and C_3 and C_4 simple decoupling capacitors: 0.01 μF or larger. As with Fig. 2.4, experiment with and without the screen and also consider enhancing the feedback capacitance by soldering short lengths of wire onto the input and output leads of the device, on the underside of the circuit board.

Fig. 3.2 Make $R_1 = 120$ k, $R_2 = 12$ k, $R_3 = 10$ k, $R_4 = 1.2$ k, $R_5 = 3.3$ k and $R_6 = 1.8$ k. Use ± 15 V supplies and decouple the power supply lines down to a ground line which should be provided on the circuit board running from one end to the other.

R_3 is only needed for protection, so we must short-circuit it to the higher frequencies, and to do this we put a 0.33 μF capacitor in parallel with R_3.

To make this circuit stable as a feedback amplifier with a gain of 100, a sensible configuration to work with, a capacitor of about 470 pF must be connected across the collector junction of Q_4, but you will need to experiment with this to get the best frequency response.

Connect the amplifier up as a feedback amplifier, as shown in Fig. 3.3(*a*), using $R_1 = 1$ k and $R_2 = 100$ k. Connect the input to a pulse or signal generator and examine the bandwidth and pulse response on both small and large signal levels.

With V_{in} (Fig. 3.3(*a*)) grounded, measure V_{out}. Hence, deduce the input bias current. Look up what is meant by 'input offset current' and 'input offset voltage', in any operational amplifier data book, and work out some way of measuring these.

This circuit should have a small signal bandwidth of over 500 kHz when connected as a feedback amplifier with a gain of 100 (40 db).

Fig. 4.11 Use 1% tolerance resistors and make $R_1 = 100\,k$, $R_2 = 500\,k$, $R_3 = 10\,\Omega$, $R_4 = 30\,k$, $R_5 = 10\,k$, $R_6 = 30\,k$, $R_7 = 6\,k$ and $R_8 = 5\,M\Omega$. C_8 is only put in to reduce the noise on the I_{in} against V_{in} display, described in Chapter 4. C_8 must not be large enough to introduce any phase shift at the measurement frequency, however.

An OP–05 is a good choice for A_1 and an OP–07 for A_2. Both should have their positive and negative supply pins decoupled to ground, as close as possible to the device, with something like 0.33 μF. This should also be done at the DUT socket.

To minimise the effects of stray capacitance across the DUT, the lead connecting the negative input of the DUT to the negative input of A_2 should be made with about 20 mm of miniature coaxial cable. The printed circuit board should then be made to have as much grounded copper as possible around the neighbourhood of the negative input of the DUT. A further improvement, for eight pin DUTs, is to use the centre eight pins of a sixteen pin socket for the DUT, and make a saw-cut all along the length of this sixteen pin socket once it has been soldered in place. A strip of copper foil may then be slipped into this saw-cut and grounded to the four pins which are not used at both ends of the sixteen pin socket. The foil acts as a screen across the DUT.

The circuit of Fig. 4.11 will not work really well in an unscreened laboratory unless it is enclosed in some kind of metal box. A simple slip-on cover is ideal because the DUT is going to be changed continuously.

Fig. 5.8 This circuit is working with very low level signals and must be enclosed in a small metal case. The photodiode should be mounted on the outside of this case, using an insulating pad of some kind, and a light-tight coupling to the light emitting diode must be arranged. This will avoid any problems with noisy ambient illumination.

Sensible values are $R_1 = 150\,k$, $R_2 = 120\,k$, $R_3 = 1.8\,k$, $R_4 = 22\,k$, $R_6 = 82\,k$, $R_7 = 15\,k$, $R_8 = 120\,\Omega$, $R_9 = 12\,k$, $R_{10} = 120\,\Omega$ and $R_{11} = 1.2\,k$. R_5 is the most convenient resistor to make variable so that the d.c. level at the output may be adjusted to zero when there is no light input. A good choice for R_5 is 270 k in series with a 20 k pre-set variable. This pre-set resistor may be put at the +15 V end of R_5 so that it may be mounted almost anywhere on the circuit board and accessed via a small screwdriver hole in the case.

C_2, C_3 and C_4, in Fig. 5.8, can all be about 0.33 μF. The 3100 should have its power supply pins decoupled to ground as close as possible to the device. The negative end of R_{11} should be decoupled to the ground at the output socket.

Fig. 6.21 This circuit has been drawn in such a way as to suggest a layout which will make each stage of the ring of five have very much the same kind of load capacitance. Unfortunately the high impedance probes which will be used to connect the oscilloscope to the circuit will upset this anyway, but at least the effect of one probe may be observed by the change in the waveform when the other probe is put on.

All the resistors in Fig. 6.21 are identical and a value of 150 k is a good choice because the speed of the circuit may then be observed over a wide range of injection current: V_+ may be taken from a few volts up to a few hundred volts without exceeding the rating of the usual 250 mW resistors.

Using general purpose devices, like 2N2222A for the even numbered transistors and 2N2905A for the odd numbered, the voltage waveforms observed at each stage are particularly interesting for the positive going transition. Saturation of the npn devices is so severe that the collector junctions stay forward biassed during the turn-on of the previous stage, and this means that the collector voltage is observed to go negative by over 200 mV before the positive going transition begins. If high speed switching transistors are used, for example 2N2369 and 2N3640, this effect is not observed and the voltage waveforms are reasonably good square waves. A further interesting idea is to use five Schottky diodes to clamp Q_2, Q_4, Q_6, Q_8 and Q_{10} in the way shown in Fig. 6.10(*a*). This will make the circuit give a good square wave with the general purpose transistors but, of course, the amplitude is about halved.

Fig. 7.10 If cheap operational amplifiers are used to make this circuit, and these have high input bias current, the resistor values must be kept down. Even so, very low frequency operation is easily obtained by using large values of capacitance. For operation close to 1 Hz, use two 1 μF capacitors in parallel for C_1 and single 1 μF capacitors for C_2 and C_3. These must, of course, be plastic film capacitors, either of good tolerance or selected by measurement. R_1 and R_2 should then be 120 k and R_3 and R_4, 100 k. The Zener diodes should be about 3 volts breakdown voltage and R_5, 10 k or less.

As the oscillator is operating at such a low frequency a very accurate measurement of the frequency may be made using classical methods, and the result compared with theory. It is interesting to observe the way in which oscillations build up in this circuit, and this may be done very simply by short-circuiting the integrator capacitor, C_3. On removing this short-circuit, oscillations should build up from the condition where all three capacitors are discharged. If the time constant associated with oscillation build up can be measured, this may be compared with the detailed theory of this third order oscillator.

Fig. 7.12 Using two BF 199 transistors, $V_+ = 10\,\text{V}$ and a 5 MHz crystal, this circuit works very well when $R_1 = 470\,\Omega$, $R_2 = 2.2\,\text{k}$, $R_3 = 680\,\Omega$, $R_5 = 330\,\Omega$, $R_7 = 680\,\Omega$, $R_8 = 39\,\Omega$ and all the capacitors are $0.1\,\mu\text{F}$. The Schottky barrier diodes should be chosen for low forward voltage and good high frequency performance. Germanium diodes can be used.

The choice of R_4 and R_6 determines the voltage across the crystal. The interesting features of the oscillator may be observed by looking at the junction of R_6 and R_7 with a high impedance probe and a wide-band oscilloscope. An FET probe is preferable because the signal is quite small. Begin experiments by making $R_6 = 39\,\Omega$ and $R_4 = 0$. The voltage waveform at the junction of R_6 and R_7 may look very square under these conditions, meaning that R_6 can be reduced. However, reducing R_6 means increasing the effective Q of the crystal, and this is why it is a good idea to begin increasing R_4. Continue reducing R_6 until the oscillator is only just oscillating. Just before this condition is reached, a good sinusoidal waveform, about 50 mV peak, should be seen at the output.

Note that the only points in this circuit which can be probed are at the emitter of Q_2, the output, and at the junction of R_6 and R_7. It is possible to confirm that the base of Q_1 is acting as a good virtual earth, however. Note the effects of small changes in V_+. The circuit should be laid out to be as compact as possible. Stray capacitance across the crystal and at the collectors of Q_1 and Q_2 should be kept as low as possible.

Figs. 8.6 and 8.7 The three parts of the system shown in Fig. 8.7, that is the two forms of Fig. 8.6 and then the three operational amplifiers A_1, A_2 and A_3, are best built on three separate boards and then connected together using a mother board with three edge connectors.

Fig. 8.6 should be built with $R_1 = 10\,\text{k}$, $R_2 = 270\,\text{k}$ and $R_3 = 27\,\text{k}$. R_4 is made about 10 k so that the output at pin 12 is not always held low by pin 3, but the wisdom of this idea could be debated. R_5 is about 10 k, and $R_6 = R_5/2$. Some offset adjustment must be added to this squaring circuit so that it gives zero output at zero input. This is easily done by arranging about $\pm 2\,\text{V}$ across two pre-set variable resistors and taking the wipers of these, via $1\,\text{M}\Omega$, to pins 16 and 10, in the case of the top Fig. 8.6 circuit block of Fig. 8.7, and to pins 16 and 9 in the other case. Use $\pm 15\,\text{V}$ supplies and decouple at the power supply pins of all devices.

For the three operational amplifiers shown in Fig. 8.7, OP–05 or OP–07 devices are a good choice, but any low input bias current, low offset amplifiers will do. Make $C_7 = 0.01\,\mu\text{F}$, $R_7 = 25\,\text{k}$, $R_8 = 100\,\text{k}$ and $R_9 = 25\,\text{k}$. It will be necessary to damp the system slightly: $4.7\,\text{M}\Omega$ in parallel with the C_7 of A_2 is a good choice for the values given so far, but variation of the damping is one of the most interesting experimental variables in this circuit.

To experiment with this system, apply a low frequency sine wave of about 100 mV to the input terminal of Fig. 8.7, and observe the output as the frequency is increased from a few tens of Hertz towards a few kHz. As Fig. 8.2 shows, the restoring force on this dynamic system is zero at zero amplitude, so that what would normally be called the resonant frequency is also zero. As the amplitude of motion increases, so does the resonant frequency. This means that, instead of the symmetrical resonance curve of a linear resonator, this system has a resonance curve which is tilted over towards the high frequencies. A large amplitude oscillation will build up only if we increase the input signal frequency. Then, at some critical 'jump frequency', the amplitude collapses and a completely different amplitude–frequency characteristic is observed when the input frequency is now reduced. A second 'jump frequency' is then observed as we jump back to the original amplitude–frequency characteristic. Work with this system is a very instructive exercise in non-linear dynamics. It is also interesting to simulate the system on a digital computer, using the same range of input frequency and amplitude which is available on the analog.

Check the amplitudes of the signals at the outputs of A_2 and A_3 during these experiments to ensure that the dynamic range of the system is not exceeded.

Observe v and dv/dt on an X,Y oscilloscope by displaying the output of A_1 as X and the output of A_2 as Y. The resulting traces, as input frequency and amplitude are varied, are most unusual and are the clearest way of observing the sub-harmonic resonances that are a characteristic of this system at high levels of drive.

Fig. 9.2 T_1 should have a 24 V rms secondary winding and the primary is best supplied from the a.c. main supply *via* a Variac. This allows us to see the effect of large changes in a.c. supply voltage on the stabiliser. For Q_1, a BC441 or 2N5320 is a good choice to work with the general purpose operational amplifier type 741 for A_1. Sensible component values are then: $R_1 = 22\,\Omega$, $R_2 = 4.7\,k$, $R_3 = R_5 = 330\,\Omega$, $R_4 = 1.2\,k$ and $C_1 = 47\,\mu F$, when a 12 V reference is used, for example a 1N5242 for D_1.

On no-load, observe how the regulator takes control as the a.c. supply voltage is increased from zero to the full supply level. This should be done first with the oscilloscope on, say, 5 V d.c./div., and then, when the stabiliser loop begins to work, on, say, 5 mV a.c./div. Noise and ripple on the output should be of the order of a few mV, compared to several volts ripple across C_1. Add a small ceramic decoupling capacitor, for example 0.33 μF, 50 V, directly across A_1 from pin 7 to pin 4, and note any change in the noise on the output.

Now load the regulator up to about 150 mA, where it will probably drop out of control because the ripple across C_1 will become so large that the level at pin 7, on A_1, will drop too close to the level at pin 6. Try to measure the true drop in output voltage as the load current increases. This is not easy, although it is easy to measure the resistance of the leads used instead!

Finally, arrange to measure the behaviour of the circuit on transient load. To do this, connect a switching transistor, in series with 150 Ω, across the output and drive this with a pulse generator. Observe V_{out} as the switching transistor goes both on and off. The transient is not the same in the two cases. Why is this? Now add a large electrolytic capacitor, say 22 μF, 50 V, directly across the output. Note the completely different behaviour on transient load. A large capacitor is usually found across the output of regulated power supplies of this classical circuit shape, but this is clearly not good practice when the circuit is viewed from the point of view expressed in connection with Fig. 9.3.

Fig. 9.6 Use a standard +15 V power supply for V_+ and work at quite low values of current so that a small general purpose switching transistor, like the 2N2369A, and diode, an 1N914, may be used for Q_1 and D_1. The two inductors can then be made quite simply, because only a few mH are needed when the square wave drive is at 50 kHz. Make $C_1 = C_2 = 1\ \mu$F, using plastic film capacitors for low inductance and series resistance. Make R_L about 100 Ω. Make T_1 a trifilar winding around several ferrite toroids: 20 + 20 + 20 turns through a stack of five T38–R10 toroids from Siemens, for example. The primary is then made by putting two of the 20 turn windings in series, while the secondary is the remaining 20 turn winding. The inductors, L_1 and L_2, need to be at least 10 mH each. These are best wound on slotted formers, 25 mm diameter and 25 mm long, having ten slots, 2 mm wide and 3 mm deep, turned into them. On these formers, a ten section winding of 80 turns in each section can be made, using enamelled copper wire, 150 μm diameter. A coupling constant of 0.5 will only just be approached when two such coils are placed end to end. Placing ferrite rods, of the kind used for antennae in a.m. radio receivers, through both coils will allow $k > 0.5$ to be achieved easily.

Fig. 9.12 This experimental circuit is stable when $R_L = 50\ \Omega$ or less. It is a good idea to commence experiments with the 50 Ω load as this keeps the dissipation levels down. Use a standard ±15 V power supply.

The output devices are VPO300M, for Q_5, and VNO300M, for Q_6. Both are in TO–237 packages having a small tab that needs a heat sink. This can be arranged very easily by clamping the tab between two plain washers on a 5 mm, or larger, nut and bolt. Use 2N2222A for Q_1 and Q_2, 2N2907A for Q_3 and Q_4, and a CA3140 for A_1.

A complete list of suitable component values would be: $R_1 = 100\,\Omega$, $R_2 = 10\,\text{k}$, $R_3 = 20\,\text{k}$, 10 turn pre-set, $R_4 = R_5 = R_{14} = R_{15} = 100\,\Omega$, $R_6 = R_{10} = 1\,\text{k}$, $R_7 = R_{11} = 680\,\Omega$ (not only does the n-channel Q_6 call for a lower V_{GS}, as shown in Fig. 9.8, it calls for less loop gain, from Q_3 and Q_4, because its g_m is higher than the g_m of Q_5), $R_8 = R_9 = 2.2\,\text{k}$, $R_{12} = R_{13} = 150\,\text{k}$, $R_{16} = R_{17} = 1\,\text{k}$ and $R_{18} = R_{19} = 4.7\,\Omega$. Large electrolytic decoupling capacitors should be connected from the source of Q_5, and the source of Q_6, to the ground end of R_L. Small ceramic capacitors, say $0.33\,\mu\text{F}$, 50 V, should be in parallel with these electrolytics, and two more $0.33\,\mu\text{F}$ capacitors should be used on A_1 from pin 7 to ground and pin 4 to ground. A good ground should run from input end to output end of the circuit board.

Check that the amplifier has an overall gain of 100, when the output across R_L is a few volts peak to peak, and that it is independent of input frequency up to several hundred kHz. Check the noise on the output when there is no input. There should be no sign of any high frequency oscillations. Check the quiescent current on zero input by measuring the d.c. drop across R_{18} and R_{19}. Zero output is, of course, first set by adjustment of R_3.

Now check the maximum output level by driving the amplifier until the output sine wave is observed to limit. This should be within about 2 volts of the supply rails. Observe this first at 1 kHz and then at high frequency.

Run the amplifier at 8 volts peak output for a few minutes. This should be close to the maximum dissipation condition. Then, to check the thermal stability of the design, remove the input and measure the d.c. levels across R_{18}, R_{19} and R_L.

Author index

Subject index

Printed in the United States
By Bookmasters